The Man Who Knew Too Much

PUBLISHED TITLES IN THE GREAT DISCOVERIES SERIES

David Foster Wallace
Everything and More: A Compact History of ∞

Sherwin B. Nuland
The Doctors' Plague: Germs, Childbed Fever, and the Strange Story of Ignác Semmelweis

Michio Kaku
Einstein's Cosmos: How Albert Einstein's Vision Transformed Our Understanding of Space and Time

Barbara Goldsmith
Obsessive Genius: The Inner World of Marie Curie

Rebecca Goldstein
Incompleteness: The Proof and Paradox of Kurt Gödel

Madison Smartt Bell
Lavoisier in the Year One: The Birth of a New Science in an Age of Revolution

George Johnson
Miss Leavitt's Stars: The Untold Story of the Forgotten Woman Who Discovered How to Measure the Universe

David Leavitt
The Man Who Knew Too Much: Alan Turing and the Invention of the Computer

FORTHCOMING TITLES

Richard Reeves on Rutherford and the Atom

Daniel Mendelsohn on Archimedes and the Science of the Ancient Greeks

David Quammen on Darwin and Evolution

William T. Vollmann
Uncentering the Earth: Copernicus and The Revolutions of the Heavenly Spheres

General Editors: Edwin Barber and Jesse Cohen

BY DAVID LEAVITT

SHORT FICTION

Family Dancing

A Place I've Never Been

Arkansas: Three Novellas

The Marble Quilt

Collected Stories

NOVELS

The Lost Language of Cranes

Equal Affections

While England Sleeps

The Page Turner

Martin Baumann; or, A Sure Thing

The Body of Jonah Boyd

NONFICTION

Italian Pleasures (with Mark Mitchell)

In Maremma: Life and a House in Southern Tuscany (with Mark Mitchell)

Florence, A Delicate Case

The Man Who Knew Too Much

DAVID LEAVITT

The Man Who Knew Too Much

Alan Turing and the Invention of the Computer

ATLAS BOOKS

W. W. NORTON & COMPANY
NEW YORK · LONDON

Printed in the United States of America
First Edition

Grateful acknowledgment is made to the following for permission to reprint previously published material: The Estate of Alan Turing: Excerpts from Alan Turing's essays, papers and letters. The University of Chicago Press: Excerpts from *Wittgenstein's Lectures on the Foundations of Mathematics, Cambridge 1939*, edited by Cora Diamond. © 1975, 1976 by Cora Diamond.

For information about permission to reproduce selections from this book, write to Permissions, W. W. Norton & Company, Inc., 500 Fifth Avenue, New York, NY 10110

Manufacturing by RR Donnelley, Bloomsburg Division
Book design by Chris Welch
Production manager: Julia Druskin

Library of Congress Cataloging-in-Publication Data

Leavitt, David, 1961–
The man who knew too much : Alan Turing and the invention of the computer / David Leavitt.
p. cm. — (Great discoveries)
"Atlas books."
Includes bibliographical references and index.
ISBN 0-393-05236-2 (hardcover)
1. Turing, Alan Mathison, 1912–1954. 2. Mathematicians—Great Britain —Biography. 3. Gay men—Legal status, laws, etc.—Great Britain. 4. Artificial intelligence—History. I. Title. II. Series.
QA29.T8L43 2005
510'.92—dc22

2005018034

Atlas Books
10 E. 53rd St., 35th Fl., New York, N.Y. 10022

W. W. Norton & Company, Inc.
500 Fifth Avenue, New York, N.Y. 10110
www.wwnorton.com

W. W. Norton & Company Ltd.
Castle House, 75/76 Wells Street, London W1T 3QT

1 2 3 4 5 6 7 8 9 0

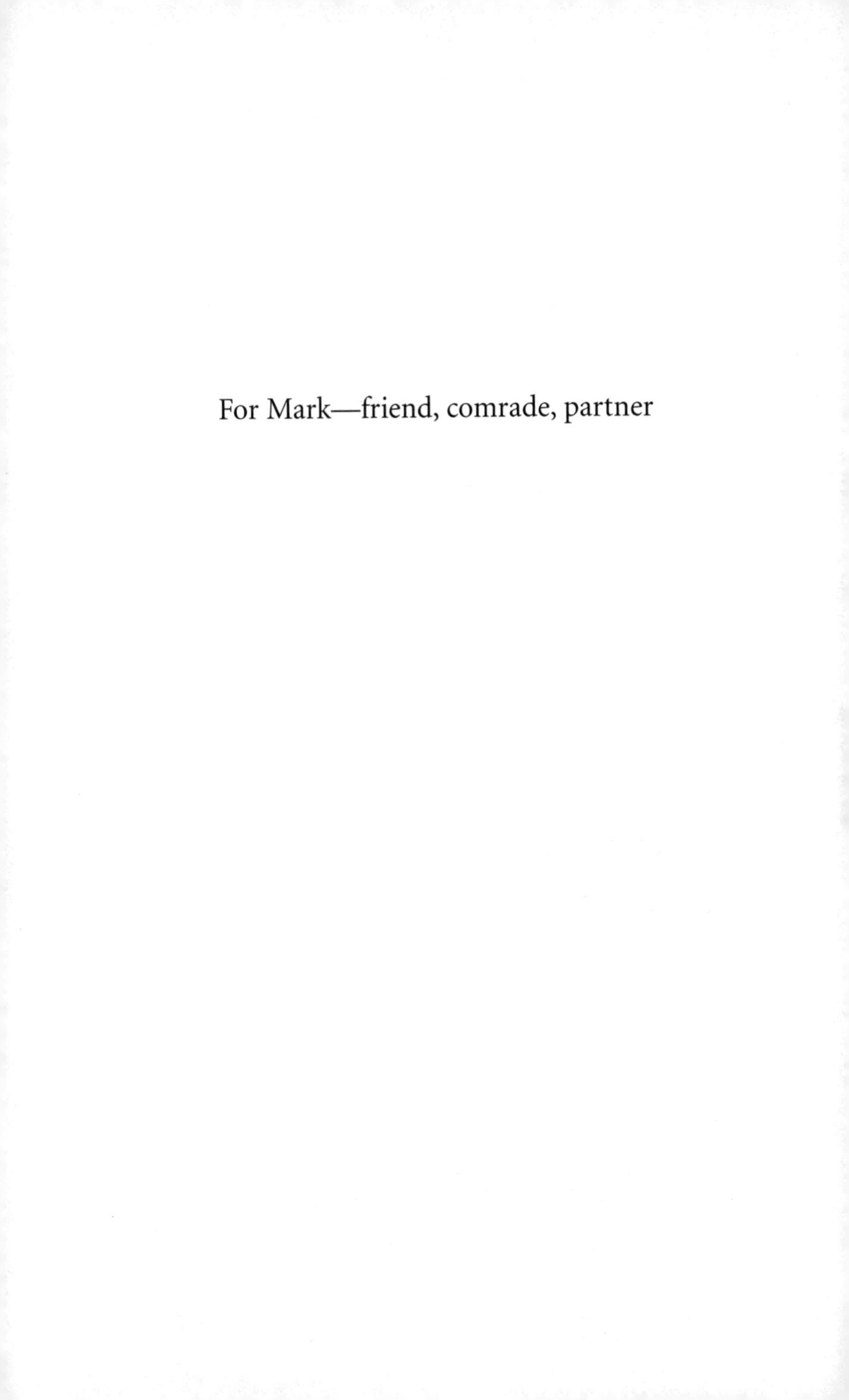

For Mark—friend, comrade, partner

Contents

The Man Who Knew Too Much

1

The Man in the White Suit

In Alexander Mackendrick's 1951 Ealing comedy *The Man in the White Suit*, Alec Guinness plays Sidney Stratton, a dithery, even childlike chemist who creates a fabric that will never wear out or get dirty. His invention is heralded as a great step forward—until the owners of the textile mills at which he was employed, along with the members of the unions representing his fellow workers, realize that it will put them all out of business. Soon enough, these perennial antagonists join forces to trap Stratton and destroy his fabric, which he is wearing in the form of a white suit. They chase him down, corner him, and seem about to murder him, when at the very last moment, the suit begins to disintegrate. Failure thus saves Stratton from the industry he threatens, and saves the industry from obsolescence.

It goes without saying that any parallel drawn between Sidney Stratton and Alan Turing—the English mathematician, inventor of the modern computer, and architect of the machine that broke the German Enigma code during World

War II—must by necessity be inexact. For one thing, such a parallel demands that we view Stratton (especially as portrayed by the gay Guinness) as at the very least a protohomosexual figure, while interpreting his hounding as a metaphor for the more generalized persecution of homosexuals in England before the 1967 decriminalization of acts of "gross indecency" between adult men. This is obviously a reading of *The Man in the White Suit* that not all of its admirers will accept, and that more than a few will protest. To draw a parallel between Sidney Stratton and Alan Turing would also require us to ignore a crucial difference between the two scientists: while Stratton is hounded *because* of his discovery, Turing was hounded *in spite* of it. Far from the failure that is Stratton's white suit, Turing's machines—both hypothetical and real—not only initiated the age of the computer but played a crucial role in the Allied victory over Germany in World War II.

Why, then, labor the comparison? Only because, in my view, *The Man in the White Suit* has so much to tell us about the determining conditions of Alan Turing's short life: homosexuality, the scientific imagination, and England in the first half of the twentieth century. Like Stratton, Turing was naïve, absent-minded, and oblivious to the forces that threatened him. Like Stratton, he worked alone. Like Stratton, he was interested in welding the theoretical to the practical, approaching mathematics from a perspective that reflected the industrial ethos of the England in which he was raised. And finally, like Stratton, Turing was "hounded out of the world" by forces that viewed him as a danger, much as the eponymous hero of E. M. Forster's *Maurice* fears that he will be "hounded out of the world" if his homosexuality is discovered. Dubbed a security risk because of his heroic work during World War II, Turing was arrested and tried a year after the opening of *The Man in the White Suit* on charges of committing acts of gross indecency with another man. As an alternative to a prison sentence, he was forced to endure a humiliating course of estrogen injections intended to "cure" him. Finally, in 1954, he committed suicide by biting into an apple dipped in cyanide—an apparent nod to the poisoned apple in one of his favorite films, the Disney version of *Snow White and the Seven Dwarfs*, and one which writers on Turing in subsequent years have made much of.

In a letter written to his friend Norman Routledge near the end of his life, Turing linked his arrest with his accomplishments in an extraordinary syllogism:

> Turing believes machines think
> Turing lies with men
> Therefore machines cannot think

His fear seems to have been that his homosexuality would be used not just against him but against his ideas. Nor was his choice of the rather antiquated biblical locution "to lie with" accidental: Turing was fully aware of the degree to which both his homosexuality and his belief in computer intelligence posed a threat to organized religion. After all, his insistence on questioning humankind's exclusive claim to the faculty of thought had brought on him a barrage of criticism in the 1940s, perhaps because his call for "fair play" to machines encoded a subtle critique of social norms that denied to another population—that of homosexual men and women—the right to a legitimate and legal existence. For Turing—remarkably, given the era in which he came of age—seems to have taken it as a given that there was nothing at all *wrong* with being homosexual; more remarkably, this conviction came to inform even some of his most arcane mathematical writings. To some extent his ability to make unexpected connections reflected the startlingly original—and at the same time startlingly literal—nature of his imagination. Yet it also owed, at least in part, to his education at Sherborne School, at King's College during the heyday of E. M. Forster and John Maynard Keynes, and at Princeton during the reign of Einstein; to his participation in Wittgenstein's famous course on the foundations of mathematics; and to his secret work for the government at Bletchley Park, where the necessity of contending with an elusive German cipher on a daily basis exercised his ingenuity and compelled him to loosen up his already limber mind.

The fallout of his arrest and suicide was that for years his contribution to the development of the modern computer was minimized and in some instances erased altogether, with John von Neumann, his teacher at Princeton, often being given credit for ideas that really originated with Turing.* Indeed, only after the declassification of documents relating to his work at Bletchley Park, and the subsequent publication of Andrew Hodges' magisterial 1983 biography, did this great thinker begin to receive his due. Now he is acknowledged as one of the most important scientists of the twentieth century. Even so, most popular accounts of his work either fail to mention his homosexuality altogether or present it as a distasteful and ultimately tragic blot on an otherwise stellar career.

*Martin Davis must be credited with setting the record straight on this account.

I first heard about Alan Turing in the mid-eighties, when he was often recalled as a sort of martyr to English intolerance. Although I had taken a basic course in calculus in high school, in college and afterward I'd made a point of avoiding mathematics. I'd made an even greater point of steering clear of computer science, even as I grew, like most Americans, increasingly dependent on computers. Then I started to read more about Turing, and to my own surprise, I found myself becoming as fascinated by the work he'd done as by the life he'd led. Within the daunting morass of Greek and German letters, logic symbols, and mathematical formulae that enwebbed the pages of his papers, there lay the prose of a speculative and philosophical writer who thought nothing of asking whether a computer could enjoy strawberries and cream, or of resolving an old and bothersome problem in logic by means of an imaginary machine writing 1's and 0's on an endless tape, or of putting the principles of pure mathematics to the practical goal of breaking a cipher.

Alan Turing bridged the gap between the delightfully useless and (for most people) remote landscape of pure mathematics and the factory world of industry in which the ability of a machine to multiply together giant prime numbers, or go through tens of thousands of possible letter substitutions in search of a match, or assist in the engineering of a bridge, meant the difference between financial success and failure, and in some cases between life and death. Yet it would be misleading to claim that Turing saw it as his duty or calling to effect such a bridging; on the contrary, the road he took from mathematical logic to machine building was an accidental one, and the only map he used was the one provided by his very particular, in some ways peculiar, in every way eccentric brain. He was the polar opposite of a company man, and had he been, in some sense, more "normal," he might never have made the advances that he did. It was his status as an outsider that allowed him to make the creative leaps that marked his career, and changed the world.

In a brief recollection published in the late fifties, Lyn Irvine, a novelist and the wife of the mathematician Max Newman, wrote of Turing, "Alan certainly had less of the eighteenth and nineteenth centuries in him than most of his contemporaries. One must go back three centuries (or *on* two perhaps) to place him. . . ." Her recognition of Turing as a figure who belonged to the past *and* the future is an insightful one, in that it emphasizes his failure to find a place for himself in the age in which he was born. "He never looked right in his clothes," she adds a few paragraphs later,

> neither in his Burberry, well-worn, dirty, and a size too small, nor when he took pains and wore a clean white shirt or his best blue tweed suit. An Alchemist's robe, or chain mail would have suited him, the first one fitting in with his abstracted manner, the second with that dark powerful head, with its chin like a ship's prow and its nose short and curved like the nose of an enquiring animal. The chain mail would have gone with his eyes too, blue to the brightness and richness of stained glass.

The alchemist took logical principles, wire, and electronic circuits, and made a machine. The knight defended the right of that machine to a future.

If only he had been able to save himself.

2

Watching the Daisies Grow

1.

He was a child of empire, and of the English middle class. His father, Julius, was in the Indian civil service, and it was in Chatrapur, near Madras, that Turing was conceived. Julius and Ethel Sara Turing then returned to England, where their second son was born on June 23, 1912, in a nursing home at Paddington. His full name was Alan Mathison Turing. According to his mother, "Alan was interested in figures—not with any mathematical association—before he could read and would study the numbers on lamp posts, etc." He also showed a fondness for inventing words: "quockling" for the noise made by seagulls fighting over food, "greasicle" for "the guttering of a candle caught in a draught," "squaddy" for squat and square. He seems to have had a hard time grasping the principle of the calendar, however, and later admitted that as a small child he was "quite unable to predict when [Christmas] would fall. I didn't even realize that it came at regular intervals."

When he was six, he was sent to a small school called

Hazelhurst. Already he had begun to show an incipient interest in science, once, again according to his mother, carefully concocting "a mixture in which the chief ingredient was pounded dock leaves for the cure of nettle stings, the formula for which he wrote down in all seriousness with a sense of its importance." He also set out to compile an "encyclopaedio" [*sic*] and at eight wrote what his mother calls "the shortest scientific work on record," *About a Microscope*, the entire text of which consisted of the line "First you must see that the lite is rite." Mrs. Turing goes on to report, rather modestly, that she herself taught him long division, noting that "as a child he always sought to know underlying principles, and apply them. Having at school learnt how to find the square root of a given number, he deduced for himself how to find the cube root." A drawing of him that she made in the spring of 1923 shows young Alan standing on the hockey field, stick in hand, bent over to gaze at some flowers—the caption reads, "Hockey, or Watching the Daisies Grow"—while a Hazelhurst end-of-term song included a couplet as indicative of his talents as of his attitude toward games:

Turing's fond of the football field
For geometric problems the touch lines yield.

In 1922 he received as a gift a book called *Natural Wonders Every Child Should Know*, by Edwin Tenney Brewster. In explaining biology, evolution, and nature, Brewster used the metaphor (very much contrary to his title) of machines. The idea that the body—and particularly the brain—could be thought of as a machine stayed with Turing and influenced the course of his future work. Brewster's book may also have

"Hockey, or Watching the Daisies Grow," drawn by Sara Turing and sent to Miss Dunwall, matron at Hazelhurst School, in the spring of 1923. (King's College, Cambridge)

jump-started his allergy to imprecision, evidenced when he complained in a letter to his brother, John, that the mathematics master at Hazelhurst had given "a *quite false impression* of what is meant by *x*." As his mother explains, the master's determination to pin *x* "down to something much too determinate and concrete for Alan's dawning logician's mind" disturbed her son at least in part because he feared that the other boys in his form might be misled.

After Hazelhurst, he was sent to Sherborne, one of the original public schools and the subject of Alec Waugh's 1917 novel *The Loom of Youth*. Like most public schools, Sherborne aspired to be what E. M. Forster called a "world in miniature," striving to invest in its students the political and social values of empire-building Britain and replicating most of its hypocrisies, prejudices, and double standards. Sexual experimentation, as well as romances between older and younger boys, figured prominently in the life of the public schools, even as their administrations decried such behavior as indecent. Indeed, in 1908 C. K. Scott-Moncrieff, later to become the first English translator of Proust's *À la recherche de temps perdu*, was expelled from his own public school, Winchester, after he published a story called "Evensong and Morwe Song" in *New Field*, the school's literary magazine. The story dealt explicitly with romantic and sexual intrigue among male students, as well as the violent reaction of the headmaster once the intrigue is exposed.

Turing's first term at Sherborne began just as the general strike of 1926 was breaking out; he had spent the summer in France, and as no trains were running, he had to bicycle the sixty miles to Sherborne from Southampton, a labor he undertook cheerfully and with no great anxiety. According to a report by his housemaster, Mr. O'Hanlon, his mathematics, in which he had started out well, was by the summer term of 1927 "not very good. He spends a great deal of time in investigations in advanced mathematics to the neglect of his elementary work." Thus he took the time to work out, entirely on his own, Gregory's series for $\tan^{-1}x$, without realizing that Gregory had beaten him to the punch by two centuries. As Mrs. Turing recalls, this discovery "was a cause of satisfaction to Alan himself. . . . On his asking if the series was correct, Colonel Randolph, his mathematics master, at first thought

Alan must have got it from a book in the Library." The colonel later told his mother that despite its originality, Turing's form master "complained that his work was so ill-presented that he ought to be sent down."

Mr. Nowell Smith, the headmaster at Sherborne, called him "the Alchemist," in part because of a report from the end of the Michaelmas term of 1927 in which O'Hanlon wrote, "No doubt he is very aggravating: and he should know by now that I don't care to find him boiling heaven knows what witches' brew by the aid of two guttering candles on a naked windowsill." According to Mrs. Turing, "Alan's only regret was that Mr. O'Hanlon had missed seeing at their height the very fine colours produced by the ignition of the vapour given off by super-heated candle grease." Had the wind blown the candle out, the result would have been, to borrow Turing's own term, a "greasicle." Of course, no one could have foreseen the ominous relevance that the term "witches' brew" would have both for Turing's life and for his death.

It was at Sherborne that he first began to display the stubborn literal-mindedness that would later get him into so much trouble even as it led to some of his most startling intellectual advances. For instance, when asked in an examination, "What is the *locus* of so and so?" (the shorthand is his mother's), instead of providing the expected proof, he simply wrote, "The *locus* is such and such." Later, when Mrs. Turing asked why he had not bothered to write out the proof, he replied that all he had been asked was "What is the *locus*?" That question he had answered. He was simply doing what he had been told.

Such episodes punctuate his life. During the Second World War, he enrolled in the infantry section of the home guard so that he could learn to shoot. Asked on a form, "Do you understand that by enrolling in the Home Guard you place yourself liable to military law?" he answered no, since he could conceive of no advantage to be gained in answering yes. He underwent the training and became a first-class marksman, his friend Peter Hilton later recalled, but as the war drew to a close, he lost interest in the home guard and stopped attending parades, at which point he was summoned before the authorities to explain his absences. Naturally the officer interviewing him reminded him that as a soldier it was his duty to attend parades, to which Turing replied, "But I am not a soldier." And he was not. Because he had answered no to the question on the form, he was in fact *not* subject to military law, and hence under no obligation to attend parades. As Andrew Hodges observes, this "Looking Glass ploy of taking instructions literally" led to a similar ruckus when Turing's identity card was found unsigned; he argued "that he had been told not to write anything on it."

Of course, from the standpoint of mathematical logic, in each of these instances Turing was behaving with utmost correctness. Mathematical logic is distinct from ordinary human discourse in that its statements both mean what they say and say what they mean, which is why a sentence like "Don't worry about picking me up, I'll just walk home through the sleet on my bad leg" is unlikely to find its way into a logic textbook. *Star Trek*'s Mr. Spock was notoriously insensitive to implication, double entendre, and passive aggression, and there was more than a touch of Mr. Spock in Turing, who often got into trouble because of his inability to "read between the lines."

All told, he did not do badly at Sherborne. He was a passable athlete, and though on one occasion he had to contend with a master who shouted, "This room smells of mathematics! Go out and fetch a disinfectant spray!" his teachers and his fellow students as a rule appreciated his talents and encouraged him in them. (The teachers, however, complained routinely that his work was untidy.) He even made a few friends, among them Victor Beuttell, whose father, Alfred Beuttell, had in 1901 invented something called the "Linolite electric strip reflector lamp." In 1927 Beuttell was at work on a new invention, the "K-ray Lighting System," which was intended to provide uniform illumination for pictures or posters. When he asked Turing to help him find a formula for determining what should be the proper curvature of the glass used, the boy not only came up with one immediately but pointed out that the thickness of the glass would also affect the illumination—something no one else had noticed. Beuttell gratefully made the necessary changes, and the lighting system was soon put into production.

A few years later, at Cambridge, Turing would give his friend Fred Clayton "the impression that public schools could be relied upon for sexual experiences." How much experience he actually had at Sherborne remains unclear, despite an ambiguous reference, in Mrs. Turing's memoir, to his having kept a "private locked diary" that another boy "out of mischief or from some other motive" stole and forced open. The unnamed perpetrator "irreparably damaged the book, in which was probably entered mathematical research. This piece of wantonness has deprived us of valuable records from which his early development might have been traced." Mrs. Turing concludes by recalling that the loss "very much distressed Alan" but does not consider what "other motive" might have been at play.

Turing's closest friend at Sherborne was Christopher Morcom, a boy, like him, prodigiously gifted in the sciences, whom he met in 1928. Their relationship blossomed along the classic trajectory of nineteenth-century "romantic friendship," marked by flurries of rhapsodic emotion—Turing wrote that he "worshipped the ground [Morcom] trod on"—but with a dose of mathematics thrown in; that is to say, when they were together, the boys were more likely to talk about relativity and the value of π—which Turing, in his spare time, had calculated to thirty-six decimal points—than about poetry. Despite their seemingly dry subject matter, these conversations hummed, at least for Turing, with poetic intensity. Ironically, a few decades earlier an American doctor had recommended the study of mathematics as a cure for homosexuality.

Christopher Morcom was probably not homosexual. Had the relationship progressed beyond Sherborne School to Cambridge, where Morcom had won the place at Trinity College that Turing coveted, it might well have come to the same end that met so many of his friendships, with the physical advance gently but firmly repulsed. But then in 1930, before he was even able to begin Trinity, Christopher Morcom died of tuberculosis. The loss devastated Turing. "I feel that I shall meet Morcom again somewhere and that there will be some work for us to do together," he wrote to his mother, "as there was for us to do here. . . . It never seems to have occurred to me to make other friends besides Morcom, he made everyone else seem so ordinary." Mrs. Turing herself might have been saying more than she realized when in a note to Morcom's mother, who had let Alan have some of Christopher's things, she wrote that her son was "treasuring with the tenderness of a woman the pencils and the beautiful star map and other souvenirs you gave him."

Not surprisingly, the loss of this beau ideal had the effect of fixing in Turing's imagination an ideal of romantic love before that ideal could have time either to sour or to transmute itself into an adult relationship. In E. M. Forster's novel *Maurice* (1914), the eponymous hero's love for Clive Durham first resolves into a sustained and presumably sustainable partnership (but, significantly, one that excludes sex, at Clive's insistence) and then dissolves into rancor when Clive decides to marry. Turing, by contrast, was never given the chance to follow his attraction to Chris Morcom through to whatever its inevitable outcome might have been. Perhaps as a result, he spent much of the rest of his short life seeking to replicate this great and unfulfilled love.

In the fall of 1931 Turing matriculated at King's College, Cambridge, where he was given rooms in Bodley's Court. At first glance King's might seem an ideal place for a young homosexual mathematician to have landed. The college was ornately beautiful, rich (thanks in part to the stewardship of the economist John Maynard Keynes), and renowned for an attitude of liberal tolerance. It had a very "gay" reputation. Forster, as infamous for his homosexuality as he was famous for his novels, lived within spitting distance of Turing's rooms. Had he been less shy, Turing might have made Forster's acquaintance and perhaps been invited to one of the evening gatherings in which the author, now getting on in years, read aloud from the manuscript of *Maurice*, which he had decided not to publish until after his death.* From Cambridge's flourishing aesthetic and philosophical circles, however, Maurice himself, in Forster's novel, feels shut out, and Turing in many ways bore a closer

*Coincidentally, P. N. Furbank, Forster's biographer, would be named Turing's literary executor.

resemblance to Maurice than to his creator. Though he lacked Maurice's blokishness, not to mention his instinct for practical life, he was, like Maurice, bourgeois and unfinished.* Also like Maurice, he felt no shame or doubt about his own homosexuality, and was even linked with another undergraduate for whom he had "longings" by some crossword puzzle clues in a King's magazine. In the novel, it is Clive, Maurice's first love and a self-proclaimed aesthete, who ends up backing away from his own homosexuality and marrying. Maurice, the more outwardly conventional of the pair, remains firm in his identity, as Turing would.

The climate for homosexual men and women in England in the 1930s was far from tolerant. "England has always been disinclined to accept human nature," Mr. Lasker-Jones, the hypnotist whom he consults in order to go straight, tells Maurice—an assertion evidenced by the Labouchere amendment of 1885, which criminalized unspecified "acts of gross indecency" between adult men in public or private and which would remain law until 1967. Under the terms of the amendment, Oscar Wilde had been arrested, tried, and sent to Reading Gaol. More recently, the withdrawal from circulation of Radclyffe Hall's lesbian novel *The Well of Loneliness* (1928) had provoked Forster to collect signatures in support of the book, which he privately loathed. (James Douglas of the *Sunday Express* had written of *The Well of Loneliness*, "I would rather give a healthy boy or a healthy girl a phial of prussic acid than this novel. Poison kills the body, but moral poison kills the soul.") Even within the protective walls of King's, to be as open about one's homosexuality as Turing was either

*In the full quotation, Clive reflects that Maurice is "bourgeois, unfinished and stupid."

insane or revolutionary. Or perhaps it was simply logical—further evidence of his literal-mindedness, his obliviousness to the vagaries of "the world." Turing neither glorified nor pathologized his own homosexuality. He simply accepted it and assumed (wrongly) that others would as well.

Despite this openness, or perhaps because of it, his experience of King's was remote from the ones described by its more luminary graduates in the many memoirs and novels they would afterward write. The college was famous for its links to Bloomsbury, to the world of the arts and theater. Although Turing went to see a production of George Bernard Shaw's play *Back to Methuselah*, however, he wasn't the sort of undergraduate likely to be invited to tea parties at which Shaw might be a guest. He was too shy to lend intellectual cachet, too awkward and ill-dressed to qualify as a beauty. Timidity probably kept him from approaching the dazzling sophisticates with whom he ate his meals, some of whom belonged to the famous university conversation society known as the Apostles. (Its members included Forster, Bertrand Russell, John Maynard Keynes, Lytton Strachey, Ludwig Wittgenstein, and Leonard Woolf.) Turing was not asked to join. Nor was he asked to join either the play-reading Ten Club or the Massinger Society, the members of which talked philosophy late into the night. Forster's Cambridge novel, *The Longest Journey* (1907), opens with a similar gathering: Ansell, Rickie, and their friends sit before a fire, arguing over whether a hypothetical cow remains in a field after her observer leaves (a variation on the old "if a tree falls in a forest" game). Their dialogue is at once flirtatious, idealistic, and rambunctiously boyish. Then Rickie says, "I think I want to talk," and tells the story of his youth. Turing, even if he had been invited to one of these meetings, would probably have been too shy to make such a claim on other people's time.

The problem was not that entrée into such circles was by definition closed to mathematicians: the number theorist G. H. Hardy (also homosexual) and "Bertie" Russell traveled in much the same milieu as Forster and Keynes. Both, however, possessed a worldliness and savoir faire that Turing could not hope to match. Instead, he stood on the sidelines, and watched, and read. Among other things, he read Samuel Butler's *Erewhon* (1871), with its warning against machines taking over the world from mankind. By nature a nonconformist, he flouted Cambridge's traditional division between aesthetes and athletes, and took up rowing. (He was in the college trial eights in 1931, 1933, and 1934.) He also took up the violin (after a fashion). He read *The New Statesman* and came under the influence of the King's economist Arthur Pigou, who, along with Keynes, advocated more equal distribution of wealth. He joined the Anti-War Council, the purpose of which was to organize chemicals and munitions workers to strike if war was declared, and gave a talk on "Mathematics and Logic" before the Moral Science Club. True, he did not travel with the Lytton Stracheys of his day, choosing instead to forge friendships (one of them sexual) with boys who, like him, were interested in the sciences, even if, unlike him, they knew how to tie their ties properly.* And

*His closest friendships were with Kenneth Harrison, Fred Clayton (who later wrote a protohomosexual novel, *The Cloven Pine*, under the pseudonym Frank Clare), and James Atkins. It was with Atkins that he had the extended, on-and-off sexual relationship, about which he had ambivalent feelings, because Atkins, in his mind, could not compare with the lost Christopher.

yet he was as much a citizen of King's as Risley, the Wilde-like pundit (modeled on Strachey) who so dazzles and intimidates Maurice. "At Trinity he would have been a lonelier figure," Hodges writes. Nor did Trinity welcome questioning as King's did. If Turing got as far as he did in mathematics, it was because, in Hodges' words, he was willing to "doubt the axioms," and that willingness was an essential part of the King's legacy.

To the extent that King's preached a philosophy, it was a creed of moral autonomy that had its origins in the philosophical writings of G. E. Moore (1873–1958) and, in particular, his *Principia Ethica* (1903). Moore's refutation of absolute idealism and advocacy of "goodness" as a simple, self-defining quality that should serve as the basis for daily conduct provided an ethical underpinning for the philosophy of the burgeoning Bloomsbury movement and put the crowd at King's at a significant remove from the English intellectual mainstream. As John Maynard Keynes would later recall, while he and his fellows accepted "Moore's religion, so to speak," they discarded his "morals." They were thus able to transform Moore's somewhat quaint utopianism into a credo of sexual and aesthetic liberation, according to which "nothing mattered except states of mind, our own and other people's of course, but chiefly our own. These states of mind were not associated with action or achievement or with consequences. They consisted in timeless, passionate states of contemplation and communion, largely unattached to 'before' and 'after.'" Keynes is notably careful to elide gender specification when he adds, "The appropriate subjects of passionate contemplation and communion were a beloved person, beauty and truth, and one's prime objects in life were love, the creation and enjoyment of aesthetic experience and the pursuit of knowledge. Of these love came a long way first."

Nor did such a philosophy exclude mathematics. Russell's influence is obvious in Keynes's assertion

> I have called this faith a religion, and some sort of relation of neo-platonism it surely was. But we should have been very angry at the time with such a suggestion. We regarded all this as entirely rational and scientific in character. Like any other branch of science, it was nothing more than the application of logic and rational analysis to the material presented as sense-data. Our apprehension of good was exactly the same as our apprehension of green, and we purported to handle it with the same logical and analytical technique which was appropriate to the latter*. . . . Russell's *Principles of Mathematics* came out in the same year as *Principia Ethica*; and the former, in spirit, furnished a method for handling the material provided by the latter.

Keynes then gives an example that is as extraordinary for its appropriation of the language of mathematical logic as for its evasion (once again) of gender:

> If A was in love with B and believed that B reciprocated his feelings, whereas in fact B did not, but was in love with C, the state of affairs was certainly not so good as it would have been if A had been right, but was it worse or better than it would have become if A discovered his mistake? If A

*Keynes may here be alluding to Hardy's observation that it "would seem, on Russell's theory, that if you can judge that Edward is the father of George, you should be equally capable of judging that Edward is the father of blue."

> was in love with B under a misapprehension as to B's qualities, was this better or worse than A's not being in love at all? If A was in love with B because A's spectacles were not strong enough to see B's complexion, did this altogether, or partly, destroy the value of A's state of mind?

Clearly this world, with its spectacled A's and good (or bad) complexioned B's, was one in which a homosexual mathematician should have thrived. Cambridge in general (and King's in particular) provided an ideal environment for intellectual and erotic experimentation, encouraging dissent while protecting the incipient dissident from the sort of violent counterreaction that his ideas and behavior might have provoked in a more public forum. None of this, in other words, was real—and as a testing ground, it allowed these young men to flex the muscles with which they would eventually challenge British complacency. "We entirely repudiated a personal liability on us to obey general rules," Keynes writes. ". . . This was a very important part of our faith, violently and aggressively held, and for the outer world it was our most obvious and dangerous characteristic." Such a philosophy jibed well with Forster's famous ethic of personal relations, which he voiced most controversially when he asserted that, given the choice between betraying his friend or his country, he hoped he would have the courage to betray his country. His was "the fearless uninfluential Cambridge that sought for reality and cared for truth," as he wrote in an introduction to *The Longest Journey*, yet it was also the Cambridge that took for granted its own elite removal from the ordinary world, and if Turing, as Hodges shows, was a less than ideal citizen of this Cambridge, it was at least in part because Sherborne's "dowdy, Spartan amateurism," as well as "its anti-intellectualism," had contributed to make him a man who "did not think of himself as placed in a superior category by virtue of his brains." One suspects that Turing would have appreciated the much more even-keeled portrait of the university that the novelist Forrest Reid provided in his 1940 memoir *Private Road*, in which he wrote plainly, "Cambridge, I cannot deny, disappointed me."

Indeed, it is in his mathematical research, more than in the record of his life, that one sees most vividly the fruits of Turing's tenure at King's. His initial work was in pure mathematics, specifically group theory. (A 1935 publication has the daunting title "Equivalence of Left and Right Almost Periodicity.") Early on, as at Sherborne, he was proving what had already been proven: "I pleased one of my lecturers rather the other day by producing a theorem," he wrote to his mother in January 1932, "which he had found had previously only been proved by one Sierpinski, using a rather difficult method. My proof is quite simple so Sierpinski is scored off." (The theorem was probably one from 1904, concerning lattice points.) A course on the methodology of science, given in 1933 by the astrophysicist Arthur Eddington (1882–1944), took him in the same direction, leading him to undertake—and find—a proof for why measurements, when plotted on a graph, tend to form the famous "bell curve" of statistics. Alas, Turing soon discovered that his result—the "central limit theorem"—had also already been proven, in 1922. His failure to check before starting off reflected, once again, both his solitariness and his tendency to be reckless. Nonetheless, he was encouraged to include the result in his dissertation, "On the Gaussian Error Function," the bulk of which he had finished by the end of 1934, and on March 16, 1935, on the basis of this dissertation, he was elected a fellow of King's College. Because he was still only twenty-two, a bit of doggerel circulated in Sherborne circles:

> Turing
> Must have been alluring
> To get made a don
> So early on.

The fellowship brought with it £300 per annum—not a lot, but enough to keep him going while he conducted his research. It was at this point that he first started to think about one of the core problems of mathematics: the *Entscheidungsproblem*, or, as it was known in English, the *decision problem.*

2.

> Turing believes machines think
> Turing lies with men
> Therefore machines do not think

When Alan Turing included this mordant syllogism in a 1952 letter to his friend Norman Routledge, he was not only alluding to the fearful possibility that his *behavior* would lead to the suppression of his *ideas*; he was also calling up—particularly through his use of the locution "to lie with"—the famed "liar's paradox." This paradox can be traced back to the fourth century BC, when the Cretan philosopher Epimenides declared,

"All Cretans are liars, as a Cretan poet has told me." Later Eubulides refined (which in mathematics often means generalized) this paradox to the statement "I am lying," and still later, in the fourteenth century AD, the French philosopher Jean Buridan refined it further by writing, "All statements on this page are false," on an otherwise blank page.

In essence, the liar's paradox works like this. Take the statement "All statements on this page are false." If this statement is true, then the one statement written on the page—"All statements on this page are false"—is false. But if it is false, then the statement written on the page must be true—and it is on the page on which all statements are false . . . and on and on.* Stoned undergraduates have for years stared up at ceilings pondering the implications of the paradox, which I first learned about in the late 1960s, from an episode of *Star Trek* called "I, Mudd." At episode's end, the eponymous villain, Harry Mudd, incapacitates a superandroid called Norman by compelling him to process a version of the liar's paradox. As Norman spits out the running loop of contradictions ("everything I say is a lie, therefore I am lying, therefore everything I say is the truth"), his speech gets faster and his voice gets higher in the manner of a tape being played at an accelerated speed. Eventually he more or less explodes, then shuts down—and that is the point. Absurd, contradictory statements are disabling. If you think about the liar's paradox too much, like Norman, it'll blow your mind.

*In *Infinity and the Mind*, Rudy Rucker describes a visit to the famous Bocca della Verità in Rome, a Roman drain cover carved to resemble a face with an open mouth; according to legend, anyone who puts his hand into the mouth and tells a lie will not be able to pull it out again. Rucker put his hand into the mouth and said, "I will not be able to pull my hand back out again."

Of course, a certain kind of astute reader who believes in the "real world" (someone, in fact, rather like Wittgenstein) will here raise an objection or two, pointing out that when I implement the liar's paradox in its most watertight form—when I say, "I am lying"—I am neither telling the truth in the sense that I tell the truth when I say, "I am writing a book about Alan Turing," nor lying in the sense that I lie when I tell my editor I'm further along on that book than I actually am. Instead, I am making an intellectual parry in an arena where statements are symbols, and where meanings matter less than the relations between them. This is the arena in which the battle to establish a solid foundation for mathematical thought has generally been fought—a battle in the course of which many luminaries have fallen. Still more mathematicians have refused to venture anywhere near the place. When I asked a Portuguese mathematician of my acquaintance whether he had any insight to offer me on the subject, he replied, "The foundations of mathematics are full of holes and I never felt comfortable dealing with such things."

Full of holes. Earlier generations of mathematicians assumed that the stability of the landscape on which mathematical structures were built was guaranteed by God or nature. They strode in like pioneers or surveyors, their task to map the fundamentals and in so doing secure the territory that future generations would colonize. But then the holes—of which the liar's paradox is merely one—started popping up, and the mathematicians started falling in. Never mind! Each hole could be plugged. But soon enough another would open, and another, and another . . .

Bertrand Russell (1872–1970) spoke for any number of idealistic mathematicians when he wrote in 1907,

> The discovery that all mathematics follows inevitably from a small collection of fundamental laws, is one which immeasurably enhances the intellectual beauty of the whole: to those who have been oppressed by the fragmentary and incomplete nature of most existing chains of deduction, this discovery comes with all the overwhelming force of a revelation: like a palace emerging from the autumn mist as the traveller ascends an Italian hill-side, the stately storeys of the mathematical edifice appear in their due order and proportion, with a new perfection in every part.

I remember that when I read George Eliot's *Middlemarch* in college, I was particularly fascinated by the character of Mr. Casaubon, whose lifework was a *Key to All Mythologies* that he could never finish. If Mr. Casaubon's *Key* was doomed to incompletion, my astute professor observed, it was at least in part because "totalizing projects," by their very nature, ramify endlessly; they cannot hope to harness the multitude of tiny details demanded by words like "all," just as they cannot hope to articulate every generalization to which their premises (in this case, the idea that all mythologies have a single key) give rise. Perhaps without realizing it, my professor was making a mathematical statement—she was asserting the existence of both the infinite and the infinitesimal—and her objections to Mr. Casaubon's *Key* hold as well for any number of attempts on the part of mathematicians to establish a *Key to All Mathematics.*

Consider, for instance, the never-written project of which G. W. Leibniz (1646–1716) dreamed at the end of the seventeenth century: to create a sort of encyclopedia comprising all human knowledge and then to translate it into mathematical symbols that could be manipulated according to rules of deduction. Leibniz called this program a *calculus ratiocinator.* "If controversies were to arise," Russell wrote (ventriloquizing Leibniz), "there would be no more need of disputation between two philosophers than between two accountants. For it would suffice to take their pens in their hands, to sit down to their desks, and to say to each other (with a friend as witness, if they liked), 'Let us calculate.'"

Doomed to failure though it was, Leibniz's "grand program" did at the very least give rise to the discipline of symbolic logic as it was later developed by George Boole (1815–1864) and Gottlob Frege (1848–1925). Boole was a schoolmaster before he became professor of mathematics at Queen's College, Cork, and perhaps for this reason his writings—principally *The Mathematical Analysis of Logic*, published in 1847—display little of Leibniz's ostentation; on the contrary, an appealing modesty and remoteness from worldly ambition (also seen in Turing) are evident in his work. In essence, Boole's objective was to establish a system for transforming logical propositions into equations. Thus, even as he employed real-world examples (white things, horned things, sheep, horned white sheep), his emphasis was in fact on the *dissociation* of the symbols he used from the situations they described; in his hands, episodes that required deductive reasoning or decision making were reduced to basic procedures in which the operative terms were "and" and "not," while the white sheep and the horned sheep were *w* and *h*.* In such a system, Boole wrote, "every process will repre-

*In "Boolean algebra," + and - give way to prepositions. A major breakthrough in computer design came with the realization that switches, which have only two states, on and off, could be made to correspond to the operations at the heart of binary arithmetic: 0 and 1, true and false.

sent deduction, every mathematical consequence will express a logical inference. The generality of the method will even permit us to express arbitrary operations of the intellect, and thus lead to the demonstration of general theorems of ordinary mathematics."

Frege* took Boole's ideas a step farther, not just by complicating them but by using them to lay the foundations for "logicism," the principal thesis of which was "that arithmetic is a branch of logic and need not borrow any ground of proof whatever from either experience or intuition." His *Begriffsschrift*, published in 1879, sought to establish "a formal language, modeled on that of arithmetic, for pure thought." With such a language, stories about the *stuff* of the world—teapots, cars, dogs, wicked queens, apples, not to mention Boole's white sheep and horned sheep—could be distilled into strings of symbols the *sense* of which was completely beside the point. Frege also provided a strict definition of mathematical proof that has not been challenged, and in his 1884 *Die Grundlagen der Arithmetik* (The Foundations of Arithmetic) took on the question of what cardinal numbers actually are,† defining each number *n* as the *class* or *set* of all collections with *n* members: "7," for example, would be defined as the set of all collections with *seven* members, everything from the

*It is probably worth noting here that Frege was not a very likable character; to quote the philosopher Michael Dummett, he was "a virulent racist, specifically an anti-semite . . . a man of extreme right-wing opinions, bitterly opposed to the parliamentary system, democrats, liberals, Catholics, the French, and, above all, Jews, who he thought ought to be deprived of political rights and, preferably, expelled from Germany."

†Cardinal numbers are the nouns of mathematics (1, 2, 3, . . .), while ordinal numbers are the adjectives (1st, 2nd, 3rd, . . .).

Seven Dwarfs to the Seven Hills of Rome to the seven letters in the word "letters." In such a system, as Russell later explained, a "particular number is not identical with any collection of terms having that number: the number 3 is not identical with the trio consisting of Brown, Jones, and Robinson. The number 3 is something which all trios have in common, and which distinguishes them from all other collections." This definition was more rigorous than those which preceded it, in that it drew a distinction between the collection itself (Brown, Jones, and Robinson) and its category (3); it also contributed significantly toward Frege's goal of constructing an axiomatic theory of arithmetic.

The first volume of Frege's magnum opus, *Die Grundgesetze der Arithmetik* (The Basic Laws of Arithmetic), was published in 1893. In contrast to the *Grundlagen*, which included no symbolism and only sketches of proofs (as opposed to proofs that would meet Frege's own rigorous standard), the *Grundgesetze* aspired to achieve the goal of using logic to establish a foundation for the practice of mathematics. But then on June 16, 1902, just as the second volume was about to go to press, Russell sent Frege a letter (in German) in which, having first praised the *Grundgesetze*, he noted, "There is just one point where I have encountered a difficulty." He then effectively undermined Frege's entire program.

The problem, in essence, had to do with the idea of *sets of sets*. Already Frege had defined the number 7 as the set of all sets with seven members: the Seven Deadly Sins, the Seven Hills of Rome, the Seven Dwarfs, etc. This set might be imagined as a box labeled "Sets with 7 Members." A similar box might be labeled "Sets with an Even Number of Members," another simply "Couples." Some sets could be members of themselves, and some could not. Consider, for instance, the set of all dogs, of which my fox terrier, Tolo, is a member. Is this set a member of itself? No: as Russell put it, mankind is not a man, just as "all dogs" is not any particular dog. Other sets, however—for example, the set consisting of "things that are not a dog"—*are* members of themselves, since whatever "a thing that is not a dog" is, it is most emphatically *not* Tolo or any other particular dog. Likewise "the set of all sets with infinite members" is a member of itself, since it has infinite members.

This was where the "difficulty" entered in. Imagine a set labeled "Sets That Are not Members of Themselves." Is this set a member of itself? If it is, then by definition it is one of the sets that are not members of themselves, in which case it is *not* a member of itself. If it is not, then it is *not* one of the sets that are not members of themselves, in which case it *is* a member of itself. Russell liked to phrase this cousin of the liar's paradox, which would come to be known as Russell's paradox or Russell's antimony, by positing a male barber who daily shaves every man in his town who does not shave himself, and no one else. If the barber does not shave himself, he is one of the men who do not shave themselves, and thus must shave himself. On the other hand, if he does shave himself, he is one of the men who do shave themselves, and therefore he must not shave himself.

Russell's letter devastated Frege, who had to hurry to insert an appendix into the second volume of the *Grundgesetze* acknowledging the contradiction (or as Russell called it, more ominously, the "Contradiction"). Naturally distraught, he replied on June 22,

> Your discovery of the contradiction caused me the greatest surprise and, I would almost say, consternation, since it has shaken the basis on which I intended to build arithmetic. . . . It is all the more serious since, with the loss of my Rule V, not only the foundations of my arithmetic, but also the sole possible foundations of arithmetic, seem to vanish.

Subsequently Frege and Russell worked together to try to resolve the paradox or, short of that, to find a means of keeping it from infecting the foundational system that they were trying to build. Frege, however, soon gave up on this ambition, focusing his attention instead on the philosophy of language, while after much effort Russell did find a rather serpentine route around the paradox he himself had brought into the world. Unfortunately, the complexities of the jerry-rigging that Russell had to perform meant that *his* magnum opus—the three-volume *Principia Mathematica*, coauthored with Alfred North Whitehead, and describing a formalized mathematical system based on a set of axioms (general propositions the truth of which is self-evident) and rules of inference through which any piece of correct mathematical reasoning could be expressed*—was both unwieldy and difficult to use.

Still, the *Principia Mathematica* did work—and well enough that in 1928, when the German mathematician David Hilbert (1862–1943) gave a famous address calling for proofs of the completeness, consistency, and decidability of mathematics, *PM*, as it was commonly called, provided the testing

*John Kemeny described *Principia Mathematica* as "a masterpiece discussed by practically every philosopher and read by practically none."

ground on which Kurt Gödel and, later, Alan Turing tried their hands. Gödel tackled completeness and consistency, Turing decidability. The results changed mathematics irrevocably—and took it in directions of which Frege had not dreamed.

3.

Hilbert's ambition was to establish and secure the foundations of formalized mathematical systems. *PM*, for all its cumbersomeness, is the classic example of such a system, in that it was designed so that from its axioms and rules of inference any true mathematical sentence could be derived. Yet Hilbert's program differed from Russell's and Frege's on two key philosophical points. First, Hilbert repudiated what Hardy called "the extreme Russellian doctrine, that all mathematics is logic and that mathematics has no fundamentals of its own," allying himself instead with Kant, who argued "that mathematics has at its disposal a content secured independently of all logic and hence can never be provided with a foundation by means of logic alone." Second, whereas Russell viewed logic and mathematics, in Hardy's words, as "substantial sciences which in some way give us information concerning the form and structure of reality" and argued that "mathematical theorems have *meanings*, which we can understand directly, and this is just what is important about them," Hilbert viewed mathematics as a *formalized* system, in which the elementary signs were drained of all meaning. Postulates and theorems would thus be regarded as strings of meaningless marks that could be put together, taken apart, and put together again in a new way simply by applying a preestablished set of rules.

Hilbert's invocation of Kant provoked skepticism in Hardy, who made rather facetious fun of his faith in "concrete signs," writing, "I had better state at once what is to me a fatal objection to this view. If Hilbert has made the Hilbert mathematics with a particular series of marks on a particular sheet of paper, and I copy them on another sheet, have I made a *new* mathematics? Surely it is the *same* mathematics and that even if he writes in pencil and I in ink, and his marks are black and mine are red . . ." For Hardy, the axioms of formalist mathematics could be likened to "the chessmen, the bat, ball and stumps, *the material with which we play*. . . . To use Weyl's illustration, we are playing chess. The *axioms* correspond to the given position of the pieces; the *process of proof* to the rules for moving them; and the *demonstrable formulae* to all possible positions which can occur in the game." But the game has no *meaning* in the sense that the king has no kingdom, the queen no lover, and the pawns no land to till; it is "cardinal in Hilbert's logic that, however the formulae of the system may have been suggested, the 'meanings' which suggested them lie entirely outside the system, so that the 'meaning' of a formula is to be forgotten immediately it is written down."

Hardy's objections notwithstanding, formalist mathematics allowed Hilbert to make an important step forward. Just as it is possible to discuss and analyze a particular game of chess, it is also possible to make general statements or judgments *about* chess. Now Hilbert showed that, by the same logic, one could make statements *about* a formalized (if meaningless) system. Such statements Hilbert defined as falling under the category of "metamathematics." Thus (to borrow an example from Ernest Nagel and John R. Newman), 2 + 3 = 5 is a mathematical expression. But the statement "2 + 3 = 5 is an arithmetical formula" belongs to metamathematics "because it characterizes a certain string of arithmetical signs as being a formula." Likewise the statement "Any formalized mathematical system is complete, consistent, and decidable" belongs to metamathematics. By *complete*, Hilbert meant that within that system, any true statement could be formally proven and any false statement formally disproven. By *consistent*, he meant that within that system, no invalid statement, such as 2 + 2 = 5 or 1 = 0, could be arrived at through a valid process of proof. Lastly, by *decidable*, he meant that within that system, there could be shown to be a "definite method" by means of which the truth or falsity of any statement might be ascertained. This last question was commonly referred to by its original German name: the *Entscheidungsproblem*, or "decision problem."

So strong was Hilbert's confidence in these assertions that when, in an address delivered in 1928 in Bologna, he asked for proofs of them, he took it for granted that the call would yield positive results. As early as 1900, in a famous speech in Paris, he had declared that the "conviction of the solvability of every mathematical problem is a powerful incentive to the worker. We hear within us the perpetual call: There is the problem. Seek its solution. You can find it by pure reason, for in mathematics there is no *ignorabimus*." In 1931 he went further, avowing in a speech occasioned by his being made an honorary citizen of his native Königsberg that "there is no such thing as an unsolvable problem." It was in this speech, after once again disparaging "the foolish *ignorabimus*," that he made his famous exhortation: "Wir müssen wissen, Wir werden wissen." ("We must know, we shall know.")

In peacetime the International Congress of Mathematicians took place at regular four-year intervals. Because of the First World War, however, there had been no congress in 1916, whereas in 1920 and 1924, because of postwar anger at German nationalism, Germany was pointedly not invited to send a delegation. In 1928 the Italian organizers of the congress did invite the Germans. This time, however, the mathematician Ludwig Bieberbach (1886–1982), working in conjunction with L. E. J. Brouwer (1881–1966), organized a boycott to protest Germany's exclusion from earlier congresses and, more generally, the Treaty of Versailles. Hilbert did not support the boycott, and wrote in response to a letter sent out by Bieberbach, "We are convinced that pursuing Herr Bieberbach's way will bring misfortune to German science and will expose us all to justifiable criticism from well disposed sides. . . . It appears under the present circumstances a command of rectitude and the most elementary courtesy to take a friendly attitude toward the Congress." In the end, Hilbert himself led a delegation of sixty-seven mathematicians to Bologna, where he underlined the pacifist theme in the speech he gave:

> Let us consider that we as mathematicians stand on the highest pinnacle of the cultivation of the exact sciences. We have no other choice than to assume this highest place, because all limits, especially national ones, are contrary to the nature of mathematics. It is a complete misunderstanding of our science to construct differences according to peoples and races, and the reasons for which this has been done are very shabby ones.
>
> Mathematics knows no races. . . . For mathematics, the whole cultural world is a single country.

For Hilbert, formalism was closely allied with pacifism. Racial and national differences were mere "suggestive meanings" from which the signs had to be liberated if peace was to be achieved and then maintained. The boundaryless landscape he described brings to mind Russell's evocation of the mathematical edifice emerging "from the autumn mist as the traveller ascends an Italian hill-side"; an ideal realm unsullied by political division. Alas, not many years later the Berlin correspondent of the *Times* of London would be reporting on a meeting of mathematicians

> at Berlin University to consider the place of their science in the Third Reich. It was stated that German mathematics would remain those of the "Faustian man," that logic alone was no sufficient basis for them, and that the Germanic intuition which had produced the concepts of infinity was superior to the logical equipment which the French and Italians had brought to bear on the subject. Mathematics was a heroic science which reduced chaos to order. National Socialism had the same task and demanded the same qualities. So the "spiritual connexion" between them and the new order was established—by a mixture of logic and intuition.

Hardy, too, took note of national differences in mathematics, remarking in his rather skeptical essay on Hilbertian proof theory, "I am primarily interested at the moment in the formalist school, first because it is perhaps the natural instinct of a mathematician (when it does not conflict with stronger desires) to be as formalistic as he can, secondly because I am sure that much too little attention has been paid to Formalism in England. . . ." English pragmatism led to a natural distrust of German formalism, the creepy impersonality of which made it as appealing to the propaganda machine of the Third Reich as to Hilbert with his dreams of a world without borders.

Indeed, it is hard not to read into Hilbert's program an attempt, through mathematics, to ward off the coming nightmare, just as it is hard not to read into Kurt Gödel's subsequent derailing of that program both the death knell of prewar idealism and the advent of a bloody, off-kilter epoch in which the prevailing metaphors would be of chaos and night, not order and morning. Like Frege and Russell before him, Hilbert hoped to establish once and for all the security of the mathematical landscape (and by extension, the security of Europe): to give a proof that for the truth or falsity of any mathematical assertion—even long unproven assertions such as Goldbach's conjecture, which holds simply that any positive integer greater than 2 is the sum of two prime numbers—there *had* to exist, somewhere, a proof. And not just any proof; on the contrary, lest some naysayer from the fringe of mathematics should balk, this proof of provability had to be "absolute," by which Hilbert meant that it should use a minimum number of principles of inference and should not rely upon the consistency of another set of axioms. Only an absolute proof would guarantee that a mathematical description was uninfected by hidden contradictions such as Russell's antimony. Already, in a lecture on Georg Cantor (1845–1918) and his revolutionary work on the infinite, Hilbert had spoken of the contradictions that "appeared, sporadically at first, then ever more severely and ominously" in mathematics, summing up,

> Let us admit that the situation in which we presently find ourselves with respect to the paradoxes is in the long run intolerable. Just think: in mathematics, this paragon of reliability and truth, the very notions and inferences, as everyone learns, teaches, and uses them, lead to absurdities. And where else would reliability and truth be found if even mathematical thinking fails us?

Yet Hilbert would not admit defeat. On the contrary, he insisted that there *had* to be

> a completely satisfactory way of escaping the paradoxes without committing treason against our science. . . .
>
> We shall carefully investigate those ways of forming notions and those modes of inference that are fruitful; we shall nurse them, support them, and make them usable, wherever there is the slightest promise of success. No one shall be able to drive us from the paradise that Cantor created for us.*

An absolute proof that mathematics was airtight would eliminate forever the risk of being expelled—mathematical Adams and Eves—from that Eden.

Viewed in the light of its imminent decimation, not to mention the decimation of Europe, Hilbert's program comes across as highly idealistic, even Platonic. At its heart, after all, is the assumption that even undiscovered proofs already exist "somewhere out there"; doubt is taken away, and the mathe-

*To this remark G. H. Hardy added wryly, "The worst that can happen is that we shall have to be a little more particular about our clothes."

matician reassured that, with enough time and hard work, he or she can lasso any beast that lurks in the metaphysical wilderness. The program was the perfect expression of Hilbert's determination to endow younger mathematicians with the will to discover, since it sought to remove from the mathematical endeavor any cause for despair or even uncertainty. Instead, it promised a way out of any maze. "Wir müssen wissen, Wir werden wissen": though the unicorn itself might not exist, somewhere in the world there had to be evidence that unicorns either were or were not and, if they *were*, that their existence could be shown by some definite method. Still, Hilbert's very language suggests at least a trace of anxiety. After all, in the Judeo-Christian universe, Edens are by nature temporary. What God gives, God can also take away. Through his reference to "paradise," Hilbert seems to be bowing, albeit subconsciously, to the knowledge that though paradise may be infinite, our stay there is decidedly finite. For a serpent lurks in the trees—the paradox.

4.

On all three counts—completeness, consistency, and decidability—Hilbert turned out to be wrong. In a 1931 paper entitled "On Formally Undecidable Propositions of *Principia Mathematica* and Related Systems," the young Austrian mathematician Kurt Gödel (1906–1978) showed incontrovertibly that mathematics as we know it cannot be used to prove itself consistent or complete. Ironically, he made his first public announcement of his findings in Königsberg in 1930—the day before David Hilbert was made an honorary citizen of the city and gave his famous address.

Gödel's method was ingenious. To begin with, he posited a system in which arithmetical formulae, theorems, and sequences could be expressed in the form of numbers. First, he wrote out the basic symbols—the alphabet—of arithmetic and assigned to each symbol a number ("not" was 1, "or" was 2, etc.). Likewise, numbers were assigned to the basic marks of punctuation, addition, and multiplication (a left parenthesis was 8, a multiplication sign was 12, etc.). Finally, numbers were given for three types of variables: *numerical* variables, which could be replaced with numerals and numerical expressions; *sentential* variables, which could be replaced with formulae; and *predicate* variables, which could be replaced with predicates. With this system, Gödel showed, it was possible to express any arithmetical sentence numerically. For example, the sentence 1 + 1 = 2 would first be rewritten in the form

$$s\,0 + s\,0 = s\,s\,0$$

with "s" in this case meaning "the immediate successor of." These signs would then be rewritten giving their numerical equivalents:

$$s\,0 + s\,0 = s\,s\,0$$

$$7\;6\;11\;7\;6\;5\;7\;7\;6$$

Successive prime numbers each raised to the power of the numbers listed above would now be multiplied together:

$$2^7 \times 3^6 \times 5^{11} \times 7^7 \times 11^6 \times 13^5 \times 17^7 \times 19^7 \times 23^6 = ?$$

The answer to this multiplication is, admittedly, a number so huge as to defy calculation. Yet here is the important point: that number, according to the fundamental theorem of arithmetic,* can be broken down *only* one way, into the prime factors listed above; it represents the unique coding of a particular equation. Thus, if one were furnished with the mystery number, it would simply be a matter of calculation to break it down into discrete units that could be translated into 1 + 1 = 2. Nor does the laboriousness of the calculation really matter, since Gödel's intent was less to offer a working model than a theoretical framework, to show that *in principle* there existed a means by which to translate arithmetical sentences into Gödel numbers, and then to translate Gödel numbers back into arithmetical sentences. True, it would take a computer to do the calculations. Though Gödel was not looking forward to the computer—at least not consciously—much in his work foreshadowed its invention.

Yet the system Gödel developed allowed him to do far more than merely code mathematical statements as numbers: it made it possible for him to invent a way of expressing metamathematical sentences *about* a formal system *within* that system. In other words, he determined a means not only to rephrase sentences like 2 + 3 = 5 as long numbers but to rephrase sentences like "2 + 3 = 5 is an arithmetical formula" as long numbers, by first reformulating the sentences into symbolic "strings" and then translating the strings into his numerical code. He wasn't doing this for his health; the point was to code one particular metamathematical sentence—the one that brought the walls tumbling down.

*The fundamental theorem of arithmetic states that any positive integer can be represented exactly one way as a product of prime numbers.

This crucial sentence reads as follows: "Formula G, for which the Gödel number is *g*, states that there is a formula with Gödel number *g* that is not provable within *PM* or any related system." Sound familiar? Paradoxes always have that hollow ring to them. Essentially Gödel was positing a formula that stated its own unprovability. If such a formula is true, then it is not provable. If it is provable, then it is not true. Yet in a *complete* mathematical system, one should be able to prove or disprove every statement made using that system, while in a *consistent* mathematical system, it should be impossible either to prove a statement that is not true or to disprove a statement that is true. Gödel had just done both. If, in other words, *PM* and its related systems—*all* related systems, in effect the whole of arithmetic—was consistent, it could not be complete. And though one could in principle add axioms to the system in order to circumscribe the inconsistency, as Russell had added axioms to *Principia Mathematica* to contend with the Contradiction, it would be a fairly simple matter to construct a meta-metamathematical Gödel sentence that showed the *new* system, with the added axioms, to be likewise incomplete and inconsistent. And then if one were to add axioms to *that* new system in order to circumscribe the inconsistency, one could construct a meta-meta-metamathematical Gödel sentence to show that this new system was *also* incomplete and inconsistent.

In effect, Gödel had proven that statements could be true without being provable. This meant that any of a host of mathematical assertions might be true but not necessarily provable. For example, Goldbach's conjecture had, by the time Gödel published his theorem, remained unproven for almost 190 years.* Mathematicians in search of solutions to this and other unsolved problems were now denied any assurance that the treasures they were hunting even existed. No longer could the words "truth" and "proof" be considered mathematically synonymous—a shattering blow to the Hilbert program.

Not surprisingly, Hilbert's first reaction to Gödel's paper was one of anger. As his biographer Constance Reid writes,

> In the highly ingenious work of Gödel, Hilbert saw, intellectually, that the goal toward which he had directed much effort since the beginning of the century . . . could not be achieved. . . . The boundless confidence in the power of human thought which had led him inexorably to this last great work of his career now made it almost impossible for him to accept Gödel's result emotionally. There was also perhaps the quite human rejection of the fact that Gödel's discovery was a verification of certain indications, the significance of which he himself had up to now refused to recognize, that the framework of formalism was not strong enough for the burden he wanted it to carry.

Rather quickly, however, Hilbert adjusted and began to make an effort to deal with the new world Gödel had ushered in, heart-

*Goldbach's conjecture—still unproven as of this writing—was the subject not just of a 1992 novel (*Uncle Petros and Goldbach's Conjecture*, by Apostolos Doxiadis) but of a marketing campaign on the part of the novel's English and American publishers, Faber & Faber and Bloomsbury USA, respectively, that offered $1,000,000 to anyone who could do what the novel's Uncle Petros could not, and construct a proof. It was not a particularly risky gamble for the publishers to take, and no one won.

ened perhaps by Gödel's own admiration for Hilbert's work, as well as by the realization "that proof theory still could be fruitfully developed without keeping to the original program."

As for Gödel, the impact of his paper would be long-lasting. Although the proof left open the possibility that some new method might be found for proving the consistency of *PM* from outside, it made absolutely clear that no such proof could be written *using* the axioms and rules of *PM*. But this rendered *PM*'s claims to absoluteness null and void. Gödel had brought to an end the age of the totalizing project, of the Casaubon-like effort to provide a key to all mathematics, and after 1931 no one would try again to write a book with a title as all-encompassing as *Principia Mathematica*.

Of Russell and Whitehead's imposing tome, Gödel wrote in 1944, "How can one expect to solve mathematical problems systematically by mere analysis of the concepts occurring if our analysis so far does not even suffice to set up the axioms?" One might have expected him to stop there. Surprisingly, however, he goes on to write,

> There is no need to give up hope. Leibniz did not in his writings about the *Characteristica universalis* speak of a utopian project; if we are to believe his words he had developed this calculus of reasoning to a large extent, but was waiting with its publication till the seed could fall on fertile ground.

Gödel then quotes Leibniz as claiming that within a period of five years "humanity would have a new kind of an instrument increasing the powers of reason far more than any optical instrument has ever aided the power of vision."

How seriously are we to take this nostalgic paean to Leibniz's ancient fantasy—a fantasy, moreover, the foundational impossibility of which Gödel himself, more than anyone else, had shown to be law? Perhaps the nightmare that the dream had turned into—the hole-riddled landscape that replaced Russell's stately Italian hillside—proved too much to bear. Gödel spent the rest of his life shuttling between the United States (chiefly Princeton) and his native Austria. His later years were marked by increasingly serious bouts of mental illness, during which he developed a terror of refrigerators and radiators, and became inordinately fond, as Turing was, of the Disney film *Snow White and the Seven Dwarfs*. His death itself befitted a career built on the exploration of paradox: convinced that unnamed strangers were trying to poison him, he refused to eat, and starved to death. Yet his proofs outlived him and affected the pursuit of pure mathematics as profoundly as Einstein's theory of relativity did the study of physics. Before his paper, mathematicians had treated logical paradoxes as holes in a landscape the fundamental soundness of which they took for granted. Such anomalies, they believed, could either be filled in or circumnavigated. But now Gödel had shown that the landscape was by its very nature unstable. Beneath the surface fault lines ran. Thanks to Gödel, paradise had been lost, and the new terrain into which mathematicians had been driven was at best inhospitable, at worst hostile.

That, at least, was the perspective of the old guard. For younger mathematicians, Gödel's work opened up the possibility of a freer, more intuitive approach to the discipline even as it closed off forever old totalizing dreams. True, he had shown that it was impossible to prove the consistency of the axioms, but as Simon Singh notes,

> this did not necessarily mean that they were inconsistent. In their hearts many mathematicians still believed that their mathematics would remain consistent, but in their minds they could not prove it. Many years later the great number theorist André Weil would say: "God exists since mathematics is consistent, and the Devil exists since we cannot prove it."

It would have alarmed Turing to learn that the question into which he was delving was so theological.

3

The Universal Machine

1.

Although it was Hilbert who first called for a solution to the *Entscheidungsproblem*, the decision problem itself dates back to the thirteenth century, when the medieval thinker Raimundus Lullus (1232–1316) conceived of a general problem-solving method that he called *ars magna*. Leibniz expanded on the work of Lullus, both by calling for the establishment of a symbolic language (the *characteristica universalis*) with which to carry out the problem solving and by drawing a distinction "between two different versions of *ars magna*. The first version, *ars inveniendi*, finds all true scientific statements. The other, *ars iudicandi*, allows one to decide whether any given scientific statement is true or not." The decision problem, as Hilbert expressed it, falls under the rubric of *ars iudicandi* and "can be sharpened to a yes/no question: Does there exist an algorithm that decides the validity of any given first-order formula?*

*In symbolic logic, first-order logic (also called the first-order predicate calculus) consists of quantified statements that begin with the so-called

Before we continue, an aside about the word "algorithm." It has an interesting history. The *American Heritage Dictionary* defines an algorithm as a "step-by-step problem-solving procedure, especially an established, recursive computational procedure for solving a problem in a finite number of steps." The word itself is derived from the name of the ninth-century Persian mathematician Muhammad ibn Mûsâ al-Khowârizmi, who around 825 AD wrote an important mathematics text, *Kitab al-jabr wa'l-muqabala.* (The word "algebra" is a derivation of *al-jabr.*) As an example of an algorithm, Roger Penrose cites Euclid's algorithm, a method for finding the highest common factor of two numbers. It works like this. Pick two numbers at random—let's say 4,782 and 1,365. What is the largest single whole number that divides into both numbers without leaving a remainder? To find this out, we first divide the larger of the two numbers by the smaller:

$$4{,}782 \div 1{,}365 = 3 \text{ with a remainder of } 687$$

We now divide the smaller of the two numbers—1,365—by the remainder, 687:

$$1{,}365 \div 687 = 1 \text{ with a remainder of } 678$$

Continuing by the same method, we get the following:

$$687 \div 678 = 1 \text{ with a remainder of } 9$$

$$678 \div 9 = 75 \text{ with a remainder of } 3$$

$$9 \div 3 = 3 \text{ with a remainder of } 0$$

existential and universal quantifiers ($\exists$ . . .) ($\forall$ . . .). The first of these translates into "There exists an object such that . . ." The second translates into "For all objects, it is the case that . . ."

Therefore 3 is the highest common factor between the two numbers.

Of course, some algorithms are vastly more complex than this one, just as some are much simpler. Adding up figures by hand, for instance, requires the use of a simple algorithm. So does determining whether a number is prime. For each algorithm, there are infinitely many possibilities, as any of an infinity of numbers is fed in. What is crucial is that the algorithmic procedure is *systematic*: that is to say, the procedure is guaranteed to arrive at an answer in a finite period of time, and within a finite number of steps. In a sense, the *Entscheidungsproblem* might be described as the quest for a sort of ur-algorithm, one by means of which the validity or provability of any argument can be determined. This was a tall order, as Hilbert himself acknowledged; indeed, he called it "the main problem of mathematical logic."

It was in his *Grundzüge der theoretischen Logik*, co-written with Wilhelm Ackermann and published in 1928, that Hilbert stated his own version of the *Entscheidungsproblem*. Here a chapter entitled "The Decision Problem" begins, "From the considerations of the preceding section, there emerges the fundamental importance of determining whether or not a given formula of the predicate calculus is universally valid."* Take Goldbach's conjecture: could an algorithm establish its universal validity (or nonvalidity), its provability (or non-provability)? "There is of course no such theorem," the ever skeptical Hardy wrote, "and this is very fortunate, since if there were we should have a mechanical set of rules for the

*Hilbert posed a version of the decision problem, however, as problem 10 from his famous 1900 lecture, the text of which is included in Jeremy J. Gray's *The Hilbert Challenge*.

solution of all mathematical problems, and our activities as mathematicians would come to an end."* Certainly a positive result would have done much to counteract the dispiriting effect (for some) of Gödel's paper, since in principle that result would amount to a fulfillment of Leibniz's idealistic notion of a *calculus ratiocinator*. Such a result, moreover, was not considered unthinkable. Writing in 1931, the mathematician Jacques Herbrand (1908–1931) noted that "although at present it seems unlikely that the decision problem can be solved, it has not yet been proved that it is impossible to do so." Such a result might even allow mathematicians to put Gödel's results aside as a kind of logical aberration along the lines of the liar's paradox. One senses a division that is almost political, with one faction viewing as a victory what the other fears will bring on the collapse of mathematical endeavor itself.

Turing was probably in neither group. His isolation (not to mention his homosexuality) disinclined him to overidentify with larger collectives. Notably he avoided, during the politically turbulent years he spent at Cambridge, acquiring a political affiliation, despite his fervent (and pragmatic) opposition to war. Along the same lines, he looked upon the *Entscheidungsproblem* as simply a question that required resolution. Perhaps because he did not come at the problem hoping for either a positive or a negative result, he was able to attack it in an entirely new way.

He was first exposed to the *Entscheidungsproblem* in 1934, when he took Professor M. H. A. "Max" Newman's course on

*However, as Prabhakar Ragde points out, "a decision procedure might not be efficient (that is, it could require millions of years on a fast computer . . .), and it almost certainly would not be illuminating."

the foundation of mathematics. Newman (1897–1984) was an avatar of the branch of mathematics called topology, which deals with the formalization of such concepts as connectedness, convergence, and continuity; with the properties of geometric figures that can be stretched without tearing. At the heart of topology is set theory, which in turn led to Hilbert, which in turn led to the questions that Hilbert had posed at the 1928 conference in Bologna. Although Gödel's 1931 paper had established that the axiomatic system embodied in *PM* was undecidable and inconsistent, the *Entscheidungsproblem*, which Newman characterized as a matter of finding a "mechanical process" for testing the validity of an assertion, remained unresolved. In a memoir written after Turing's death, Newman summed up the situation at the point when Turing elected to take on Hilbert's final challenge:

> The Hilbert decision-programme of the 1920's and 30's had for its objective the discovery of a general process, applicable to any mathematical theorem expressed in fully symbolical form, for deciding the truth or falsehood of the theorem. A first blow was dealt at the prospects of finding this new philosopher's stone by Gödel's incompleteness theorem (1931), which made it clear that the truth or falsehood of *A* could not be equated to provability of *A* or not-*A* in any finitely based logic, chosen once for all; but there still remained in principle the possibility of finding a mechanical process for deciding whether *A*, or not-*A*, or neither, was formally provable in a given system. Many were convinced that no such process was possible, but Turing set out to demonstrate the impossibility rigorously.

The summer after he was elected a fellow of King's, Turing took to running long distances in and around Cambridge. His friend Robin Gandy later wrote, "I remember Turing telling me that the 'main idea for the paper' came to him when he was lying in the grass in Grantchester meadows." Gandy speculated that by then Turing had "already conceived of some form of Turing machine, and that what he meant by 'the main idea' was the realisation that there could be a universal machine and that this could permit a diagonal argument." Sometime later Turing shared this idea with his friend David Champernowne. He did not mention it to Newman, to whom he presented a finished typescript in April 1936. Just as inspiration—the rare thrill of seeing a way through—had come to him in solitude, it was in solitude that he undertook the labor of constructing and writing up the proof.

What he produced was remarkable. Earlier, Hardy had dismissed those naïve enough to assume that mathematicians made their discoveries by turning the handle of some "miraculous machine." Remember, however, that Turing was famously literal-minded. When Newman, in his lecture, described Hilbert's "definite method" as a "mechanical process," he started an idea in Turing's head the future repercussions of which would be immense. The word "mechanical," in its original sense, had referred to manual occupation, to work performed by human beings. By the 1930s, however, mechanical meant gears, rotors, vacuum tubes. It meant a machine. Turing took both definitions to heart.

In the 1930s, when he began his work on the *Entscheidungsproblem*, the word "computer" also had a meaning different from the one it has today: it meant simply a person who did computations—that is to say, a person engaged in the active use of algorithms. Computation in the 1930s required long hours of human labor, during which the computer might be aided by such tools as an abacus or even an adding machine but was nonetheless required to do the work herself.* No computational machines existed, and though the eccentric genius Charles Babbage had in the nineteenth century conceived of and designed one, his "analytical engine" was never built. Babbage's invention foreshadowed Turing's "universal machine" in that it would in principle have been capable of any mathematical calculation. It differed in that Babbage failed to make the crucial conceptual breakthrough of recognizing that the instructions could be written in the same mathematical language as the procedure to which they applied. Instead, he envisioned an essentially industrial device the basis for which was a machine designed to weave the rich patterns on brocade, with the instructions encoded on punch cards. Once again, in the case of Babbage, the milieu of computer science brushed up against that of literature, since one of his champions was Ada, the countess of Lovelace, the daughter of Lord Byron. Indeed, of Babbage's engine, Ada Byron wrote, "We may say most aptly, that the Analytic Engine *weaves algebraic patterns* just as the jacquard-loom weaves flowers and leaves."†

*I follow Martin Davis's lead in referring to the computer as "she," since, as Davis points out, at that period most "computers" were, in fact, women.

†This passage is quoted by Robin Gandy in his fascinating essay "The Confluence of Ideas in 1936." Here Gandy also mentions two little-known inventors who made proposals for universal calculating machines after Babbage: P. E. Ludgate in 1909 and L. Torres y Quevedo in 1914. Others referred to Babbage in works on building more practical machines, but in each of these cases "the emphasis is on programming a fixed iterable

According to Gandy, Turing was unaware of Babbage's planned machine when he undertook his work on the *Entscheidungsproblem.* Nonetheless, he shared with Babbage an approach that reflected the essentially industrial ethos of the England in which he had grown up. Technology, for Turing, meant factories teeming with human labor—a milieu not unlike the one which Sidney Stratton makes his discoveries in *The Man in the White Suit.* The machine of which he conceived bore a closer resemblance to a knitting or packing machine than to an iPod, though with the advent of electronics, this too would change.

Turing presented his results in a paper modestly entitled "On Computable Numbers, with an Application to the *Entscheidungsproblem.*" He finished the first draft in April of 1936, and the paper was published in early 1937 in *Proceedings of the London Mathematical Society.* It divides into roughly three sections: the first defines the idea of the "computable number" and of the "computing machine"; the second posits the concept of a "universal machine"; and the third employs these concepts to prove that the *Entscheidungsproblem* is insoluble. Like most of Turing's papers, "Computable Numbers" is marked by a curious blend of humbly phrased, somewhat philosophical speculation and highly technical mathematics. The result, for the general reader, is disconcerting, since invariably those passages the import of which is easy to grasp segue immediately into dense bogs of unfamiliar symbols, German and Greek letters, and binary numbers. Yet what is perhaps even more striking than the paper's style is its utter lack of

---

sequence of arithmetical operations. The fundamental importance of conditional iteration and conditional transfer for a general theory of calculating machines is not recognized. . . ."

intellectual grandstanding. Indeed, one comes away from reading it with the distinct sense that Turing had no clue as to the importance of what he had just done.

As is so often the case in mathematics, the central question that the paper addresses appears, on the surface at least, to be bewilderingly simple. "What," Turing asks, "are the possible processes which can be carried out in computing a number?" Already he had defined computable numbers

> as the real numbers whose expressions as a decimal are calculable by finite means. Although the subject of this paper is ostensibly the computable *numbers*, it is almost equally easy to define and investigate computable functions of an integral variable or a real or computable variable, computable predicates, and so forth. The fundamental problems involved are, however, the same in each case, and I have chosen the computable numbers for explicit treatment as involving the least cumbrous technique.

As Hodges observes, "it is characteristic of Turing that he refreshed Hilbert's question by casting it in terms not of proofs, but of computing *numbers*. The reformulation staked a clearer claim to have found an idea central to mathematics." At the same time, Turing wants to make sure that we remember (to quote Roger Penrose) that "the issue of computability is an important one generally in mathematics. . . . One can have Turing machines which operate directly on *mathematical formulae*, such as algebraic or trigonometric expressions, for example, or which carry through the formal manipulations of calculus." Such machines are more technically complex versions of the number-oriented machine that Turing posits a few sentences later: "According to my definition, a number is computable if its decimal can be written down by a machine."

The significance of this statement should not be underestimated. To speak of a hypothetical computing "machine," especially in a mathematics paper in the 1930s, was to break the rules of a fairly rigid orthodoxy. No such machines existed at the time, only calculating devices too crude to undertake any complex mathematics, and certainly not programmable. Yet Turing offers the sentence with no fanfare whatsoever, and then—just as quickly as he has brought up the important concept of the computing machine—he drops it, in order to give an outline of what the rest of the paper will contain. He returns to his machine only in the second paragraph of the next section, in which he compares "a man in the process of computing a real number to a machine which is only capable of a finite number of conditions." These conditions he calls *m*-configurations.

Turing now describes how the machine actually works. Running through it is a tape divided into squares each of which can be marked with a symbol. At any moment, only one square can be "in the machine." This square is the "scanned square," while the symbol it bears is the "scanned symbol." The scanned symbol "is the only one of which the machine is, so to speak, 'directly aware.' However, by altering its *m*-configuration, the machine can effectively remember some of the symbols which it has 'seen' (scanned) previously." The machine's behavior at any moment is determined by its *m*-configuration and by the scanned symbol, which, taken together, Turing defines as the machine's *configuration*. Depending on its configuration, the machine will write a symbol in a blank square; erase a symbol already written there;

move the tape one space to the left; or move the tape one space to the right. What determines how it will act is a "table of behavior" specifying the sequence of *m*-configurations according to which the machine can carry out its particular algorithm. "At any stage of the motion of the machine," Turing continues, "the number of the scanned square, the complete sequence of all symbols on the tape, and the *m*-configuration will be said to describe the *complete configuration* at that stage. The changes of the machine and tape between successive complete configurations will be called the *moves* of the machine." The distinction between the machine's *m*-configuration, configuration, and complete configuration is worth taking note of, because it will become relevant as the argument progresses. Although such a machine is now commonly referred to as a "Turing machine," Turing himself called it an "automatic machine" or "*a*-machine."

Before we look at an example of how a specific Turing machine does its job, it is worth reminding ourselves that when Turing wrote his paper, he was not, in fact, thinking of a machine that would or could ever be built. Nor did he share, at this stage, Babbage's Rube Goldberg–like enthusiasm for cranks and gears.* The engineer in Turing would emerge later; when he wrote "Computable Numbers," he intended his machine as a kind of literary device—the analogy, as it were, by means of which he could convey the central concept of the

*In the obituary for Turing in the *Times* of London, however, Newman wrote, "The description that he then gave of a 'universal' computing machine was entirely theoretical in purpose, but Turing's strong interest in all kinds of practical experiment made him even then interested in the possibility of actually constructing a machine on these lines."

computable numbers most cleanly and economically. Analogy is an important tool in making mathematics comprehensible to the nonmathematician; Turing was unusual in that he built the analogy into his proof. By doing so, he distinguished himself from mathematicians who were following less elegant (or, as he might have put it, "more cumbrous") avenues in their pursuit of the same idea.

So—let us return to our human computer. In keeping with the industrial ethos of the age, let us imagine her passing her days in a factory or sweatshop akin to the vast tracts of human toil depicted in Dickens novels or in *The Man in the White Suit*. This factory, however, produces not buttons or button hooks or even train engine parts, but numbers. In it, multitudes of women sit at tables, each slaving away at a different algorithm. There are as many women in the factory as there are algorithms. One is divining cube roots. Another is breaking compound numbers into primes. Still another is compiling tables of logarithms. Our particular computer, luckily or not, is responsible for one of the simplest, if least challenging, of algorithms: she is adding numbers together. Because this is a Turing factory, the computers are working not on two-dimensional pads but on one-dimensional tapes, of which, presumably, each has an infinite supply. These tapes are "divided into squares like a child's arithmetic book." This simply means that the operations most of us carry out vertically she carries out horizontally. At present our particular computer is sitting before her tape with a pencil, working out a sum:

$$\begin{array}{r} 9{,}251{,}754{,}803 \\ +\quad 746{,}380 \\ = \underline{\qquad\qquad\qquad} \end{array}$$

Or, as it would appear on the tape:

7	4	6	3	8	0	+	9	2	5	1	7	5	4	8	0	3	=

How long would it take most of us to perform this elementary algorithm? Assuming we did not use calculators, but simply lined the numbers up and performed the usual operations of adding and carrying, probably about half a minute. We might well, however, make mistakes. More importantly, in order to complete the task at hand, we would need to divide the operation up into what computer programmers call subroutines, for the simple reason that few of us have memories capable of holding all the figures at one time. Turing acknowledges this from the very beginning, noting that the justification for his definition of computable numbers "lies in the fact that the human memory is necessarily limited." This is especially the case when it comes to what Turing calls "compound symbols," of which he notes, "The difference from our point of view between the single and compound symbols is that the compound symbols, if they are too lengthy, cannot be observed at one glance. This is in accordance with experience. We cannot tell at a glance whether 9999999999999999 and 999999999999999 are the same."

It is now that Turing introduces into his argument what he calls the computer's "state of mind" as she does her work. Arguments of this sort, he freely admits, "are bound to be, fundamentally, appeals to intuition, and for this reason rather unsatisfactory mathematically. The real question is 'What are the possible processes which can be carried out in computing a number?'"

He attacks this question first by presenting a thorough account of just what is going on in the computer's head as she does her job. At any moment, he explains, her behavior is determined by two factors: the symbols she is looking at and her "state of mind." Obviously there is a limit to how many symbols she can take in at any one moment. For example, if you show me the number 352, I can remember it and repeat it back to you without difficulty. On the other hand, if you show me the number 352,798,634,001, I will probably have to break it down into discrete units in order to repeat it back. As Turing explains, if our computer wishes to observe more symbols than her memory allows, she "must use successive observations. We will also suppose that the number of states of mind which need to be taken into account is finite."

In effect, Turing is trying to break down the process of elementary arithmetic into its most basic parts, much as an inquisitive child might disassemble a machine in order to see how it works. In order to fulfill her task, he writes, the computer must first "split up" the algorithm that she is performing "into 'simple operations' which are so elementary that it is not easy to imagine them further divided." A "simple operation" is one in which "not more than one symbol is altered. Any other changes can be split up into simple operations of this kind."

To get an idea of what Turing means here, let us return to our computer and her tape. Before her she reads an equation. (Let's make it a shorter one, because the pages of a book make it difficult to print a very long segment of tape.)

	6	4	3	9	+	8	1	5	=								

The first numeral she looks at is the last of the three numerals making up the number 815: 5. She then looks at the 9 that is the last numeral of the first number, 6,439. Putting these together, she gets 14. But in order to perform this calculation—and this is Turing's crucial point—she has only to observe one square at a time. She also has to carry the 1 in 14, and in order to do this, she must insert into her calculation a numeral that will not be part of the final result but that occupies a square of the tape (and her mind) only until she is ready to add together the numerals to the left of the two she has just dealt with. In this diagram I shall indicate this temporary symbol by printing it in boldface italics:

	6	4	3	9	+	8	1	5	=				***1***	4			

The next observed squares would be the 3 from the first number and the 1 from the second. These add up to 4, to which the computer also has to add the 1 that is the remainder of the earlier addition of 9 and 5. The 1 is thus erased and replaced by the next permanent number, which is a 5.

	6	4	3	9	+	8	1	5	=			5	4				

The computer continues in this manner until she obtains the required answer:

	6	4	3	9	+	8	1	5	=	7	2	5	4				

There are two other aspects of the computational procedure that must be addressed if we are to establish a full-fledged technical description of it. The first is the problem of what Turing calls "changes of distribution of the observed squares." What if our computer is at work on an especially arduous calculation—one in which each of the figures to be added contained, say, 100 numerals? In such a case, she must observe and absorb a very long sequence of squares on the tape. But because she can take in, in a single glance, only a specific length of the tape, she will have to divide up the computational process into subprocesses.

The second matter that must be taken into account is the question of what Turing calls "immediate recognisability." Squares "marked by special symbols," one might think, would be by definition "immediately recognisable" to the computer. This makes obvious sense: when calculating, our minds will of course instantly distinguish such symbols as +, –, =, π, and ⊇ from the numbers that surround them. But what happens when we encounter a *sequence* that constitutes a special symbol, such as the numbers that in mathematical papers are used to label equations and theorems? Just as compound symbols such as 999999999999999 can be used to indicate large numbers, in "most mathematical papers the equations and theorems are numbered. Normally the numbers do not go beyond (say) 1,000. But if the paper was very long, we might reach Theorem 157767733443477; then, further on in the paper, we might find '. . . hence (applying Theorem 157767733443477) we have . . .' " But just as a human computer could compare the two numbers "figure by figure, possibly ticking the figures off in pencil to make sure of their not being counted twice," it is possible to design a machine capable of doing, more or less, the same thing.

We now have a sort of road map of the procedure by which the computer fulfills her job at the factory: through a complex and, yes, cumbrous succession of steps—observations, recognitions, operations—she is able to do the sum with which we began: she adds 746,380 and 9,251,754,803 and obtains 9,252,501,183. The "simple operations" into which her procedure breaks down consist of:

> *(a)* Changes of the symbol on one of the observed squares.
> *(b)* Changes of one of the squares observed to another square within [a certain number of] squares of one of the previously observed squares.

And since "some of these changes necessarily involve a change of state of mind," the most

> general single operation must therefore be taken to be one of the following:
>
> *(A)* A possible change *(a)* of symbol together with a possible change of state of mind.
> *(B)* A possible change *(b)* of observed squares, together with a possible change of state of mind.

"The operation actually performed," Turing concludes, "is determined . . . by the state of mind of the computer and the observed symbols. In particular, they determine the state of mind of the computer after the operation is carried out."

It is at this point that Turing offers the closest thing to a triumphant rhetorical flourish in the whole of his paper. "We may now," he writes, "construct a machine to do the work of this computer." And yet, lest his reference to "states of mind" should provoke balking from mathematicians made uncomfortable by such unorthodox methods, he first offers an alternative argument, which depends on the idea of a "note of instructions" provided to the computer before she begins her work. According to this argument, the computer—and who can blame her, given the tedious nature of her work?—takes a lot of breaks. She performs one step in her addition, then gets up and has a sandwich. She performs another step, then has a cup of tea. She performs another step, then goes to the bathroom. At this rate it's going to take her an inordinately long time to finish her work, but no matter. Speed, in this factory, is not of the essence. Besides, she has her "note of instructions"—which, of course, corresponds to the "table of behavior" with which Turing began his discussion of the hypothetical *a*-machine.

2.

So now let us make the cinematic jump that lies at the heart of Turing's paper. All at once the millions of women in the factory disappear. Our computer disappears, in order, we hope, to take up a more restful and satisfying occupation. In her place, suddenly, there sits a Turing machine. Indeed, in the place of every one of the women, there sits a Turing machine. A factory full of Turing machines, each one performing some specific algorithm. Let us for the moment focus on the machine that has replaced our particular friend, the woman who tots up figures. To simplify matters—and because the machine has just started its job—we will give it an even simpler task to perform than the ones at which its human counterpart was at work. We are going to ask the Turing machine to add 2 and 2.*

At this point we need to change the notational system that

*This example is based on one given by Andrew Hodges in the biography.

we are using. Our human computer has used the familiar Arabic notation, a system which, because it employs ten symbols (0, 1, 2, 3, 4, 5, 6, 7, 8, 9), is known as the *denary* system. In undertaking the addition algorithm, on the other hand, our machine will employ the much simpler *unary* system, which requires only one symbol: a 1. In the unary system, the number 2 is written as 11, the number 3 as 111, the number 4 as 1111, and so on. Most of us actually do make use of the unary system on occasion, when, for instance, we are keeping score during card games.

So—the tape runs through the machine. For such an elementary operation as addition, only one symbol is required: the symbol 1. As the tape approaches the machine, which is presumably situated somewhere to the left of the two sets of numbers, it reads as follows:

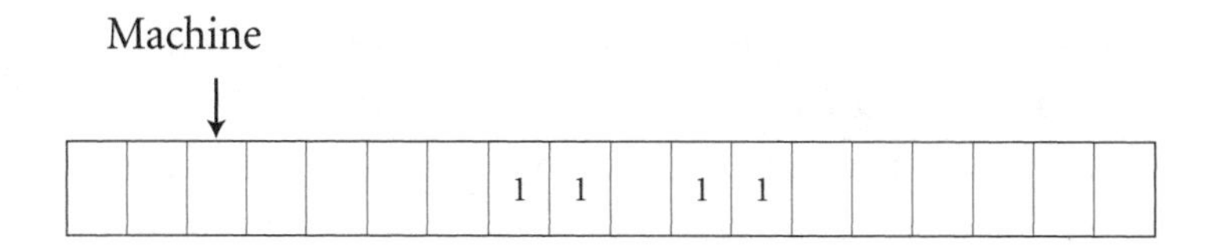

The machine works in exactly the same way as our human computer: that is to say, it observes the squares one at a time and then performs a specific operation on each of them. The operation is in every case determined by the machine's table of behavior, which in this case is set up as a series of *m*-configurations each labeled with a letter. Depending on its *m*-configuration, the machine will respond in different ways to each of the two symbols, a 1 and a blank, then move into a new *m*-configuration. (From here on I shall intermittently use "state" as a synonym for "*m*-configuration.")

m-Configuration	Symbol	ACTION	New *m*-Configuration
A	blank	Move 1 square to the right	A
A	1	Move 1 square to the right	B
B	1	Move 1 square to the right	B
B	blank	Print 1; move 1 square to the right	C
C	1	Move 1 square to the right	D
C	blank	Move 1 square to the left	D
D	blank	No move; machine stops	D
D	1	Erase 1; machine stops	D

In adding 2 and 2 (or, as the machine sees it, 11 and 11), the machine follows the instruction table, with the following result. It begins in state A, then works through the succession of blanks, to which it makes no changes. Upon arriving at the first 1, it switches into state B. In state B, it arrives at the second 1, and moves, once again, one square to the right. (For 1's, states A and B are identical.) It then arrives, in state B, at a blank, which according to its instructions it erases and replaces with a 1. The machine now shifts into state C, in which it encounters—but makes no changes to—the next 1. However, it has now shifted into state D. This means that it erases the next 1, leaving a blank. At this point the machine stops, having effected its operation. (In the following diagrams, the arrow indicates the position of the scanner on the machine.)

Machine in state A, before performing the algorithm:

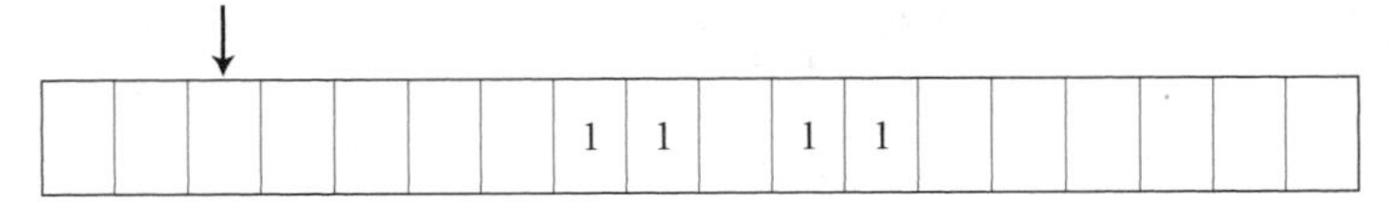

Machine in state D, after performing the algorithm:

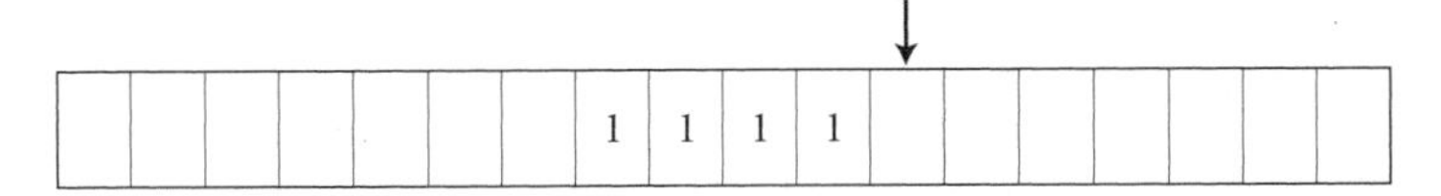

As we can see, the two sets of 2 have now been replaced by one set of 4. By means of an algorithm that never even requires the mention of the usual terms associated with addition, we have established that 2 + 2 = 4.

In "Computable Numbers," Turing gives two examples of *a*-machines. These, too, require us to make use of a new notational system—the *binary* system. To understand the difference between the binary, the unary, and the denary systems, it is helpful to imagine a ladder with infinite rungs, each of which corresponds to a natural number. In the unary system, these rungs are of equal width. In the denary system, the rungs thicken with each power of 10. That is to say, the first 9 rungs are of equal thickness. At the 10th rung a thickening occurs. The rungs remain the same thickness as the 10th rung until the 100th rung is reached, at which point the rungs thicken again, remaining the same thickness until the 1,000th rung is reached . . . and on and on. Each time one of these thickenings takes place, an extra numeral is added.

In the binary system, the rungs of the ladder thicken according to exactly the same scheme as in the denary system, except that instead of thickening with each power of 10, they thicken with each power of 2. Likewise, just as an extra numeral is added in the denary system at 10, 100, 1,000, etc., in the binary system an extra numeral is added at 2, 4, 8, 16, etc. Because the shifts occur at multiples of 2, however, the binary symbol requires only two symbols to the denary system's 10: 0 and 1.

Denary System	**Binary System**
0	0
1	1
2	10
3	11
4	100
5	101
6	110
7	111
8	1000
9	1001
10	1010
11	1011
12	1100
13	1101
14	1110
15	1111
16	10000
17	10001
18	10010
32	100000
33	100001

The great advantage of the binary system in computer programming is that it allows the use of Boolean algebra, with the 1 and 0 corresponding to the on and off position of a valve, switch, or circuit. In addition, the system permits a far more economical coding of large numbers than would be possible with the unary system. Lastly, the binary system simplifies the coding of letters, mathematical symbols, and punctuation marks, which can also be represented in binary form.

The first of the *a*-machines that Turing gives as examples is a very simple machine designed to generate the infinite sequence 010101 . . . (in this case the ellipses indicate that the sequence goes on forever without changing). This machine differs from the one described previously in that it must print and recognize *two* symbols: 0 and 1. Its table of behavior looks like this:

***m*-Configuration**	**Symbol**	**Action**	**New *m*-Configuration**
A	blank	Print 0; move 1 square to the right	B
B	blank	Move 1 square to the right	C
C	blank	Print 1; move 1 square to the right	D
D	blank	Move 1 square to the right	A

A slightly more complicated Turing machine prints the sequence 001011011101111011111. . . . This machine must be capable of printing ə, x, 0, and 1. (The odd symbol ə is used as a placeholder that indicates the initiation of the sequence, and is not a part of the sequence itself; like the x, it serves to minimize the number of steps the computer needs to take in order to get the result.) As Turing explains,

> The first three symbols on the tape will be "əə0"; the other figures follow on alternate squares. On the intermediate squares we never print anything but "x." These letters serve to "keep the place" for us and are erased when we have finished with them. We also arrange that in the sequence of figures on alternate squares there should be no blanks.*

The machine thus "scribbles" intermediate steps in the calculation much as our human computer did. Its table of behavior, needless to say, is somewhat more complicated than those for the previous machines. (Because this machine requires so many more moves, here we shall use L to signify "move one square to the left," R to signify "move one square to the right," P to signify "print," and E to signify "erase.")

m-Configuration	Symbol	Action	New *m*-Configuration
A	blank	P ə; R; P ə; R; P0; R; R; P0; L; L	B
B	1	R; Px; L; L; L	B
B	0	no action	C
C	0 or 1	R; R	C
C	blank	P1; L	D
D	x	Ex; R	C

*I have altered slightly the symbols that Turing originally used.

D	ə	R	E
D	blank	L; L	D
E	0, 1, x, or ə	R; R	E
E	blank	P0; L; L	B

Turing goes on to show the first few sequences of symbols that the table generates, along with the *m*-configurations that come into play at each stage. The tape begins in state A and prints this sequence:

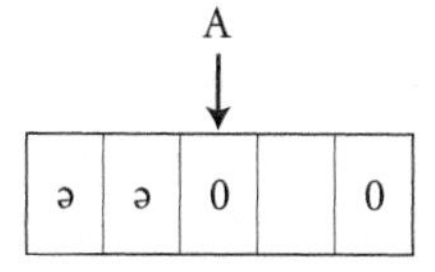

The machine, which is now scanning an 0, then shifts into state B. Following its new set of instructions, it does nothing but shifts into state C, at which point it is issued another set of instructions, telling it to move two spaces to the right. Because the machine is still in state C, and has encountered another 0, it moves two further spaces to the right:

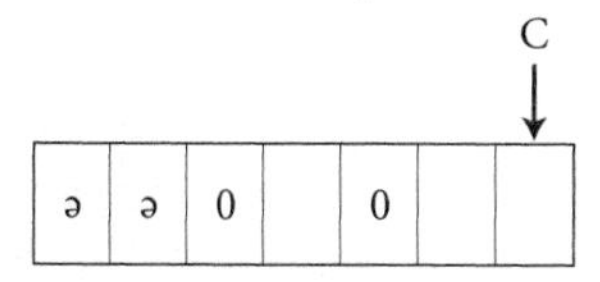

The machine has now encountered a blank square. Consulting its instruction table, it sees that when in state C, it should, upon encountering a blank square, print a 1, move one square to the left, and shift into state D:

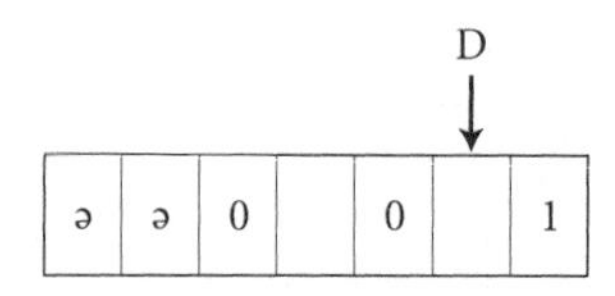

In state D, its instruction when encountering a blank square is to move two squares to the left, remaining in state D. Encountering another blank square, the machine moves two further squares to the left, where it encounters an ə. Its new instruction is to move one square to the right and shift into state E. It now scans a 0. In state E, its instruction for any symbol is to move two squares farther to the right, remaining in state E. Scanning yet another 0, it moves two *further* squares to the right, where it encounters a 1. Moving two further squares to the right, it encounters a blank. In state E, its instruction upon encountering a blank is to print a 0, move two squares to the left (where it encounters a 1), and return to state B. Following the instructions from state B for what to do when landing on a 1, the machine moves one square to the right, prints an x, then moves three squares to the left:

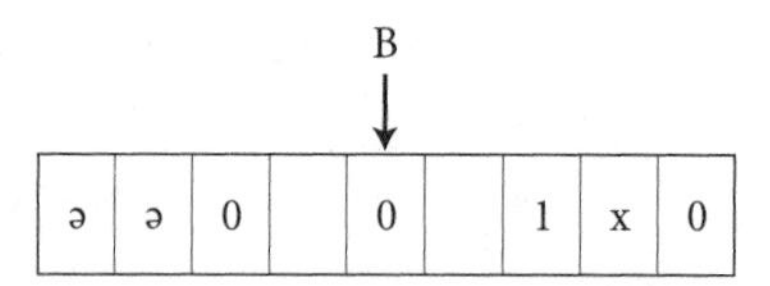

The machine, still in state B, has now met up with another 0. It shifts into state C, moves two squares to the right, encounters a 1, moves two further squares to the right, encounters a 0, moves two *further* squares to the right, encounters a blank, prints a 1, and moves one square to the left. Now the tape reads like this:

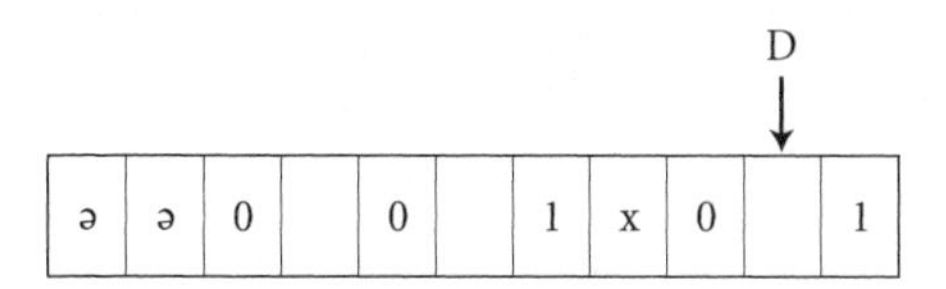

The machine is now scanning the blank just to the left of the last 1, and has shifted into state D. It moves two squares to the left, then, remaining in state D, erases the x per its instructions, and moves one square to the right, where it scans a 0. Now it shifts into state C, moves two squares to the right, scans a 1, moves two further squares to the right, scans a blank, prints a 1, moves one square to the left, and shifts again into state D:

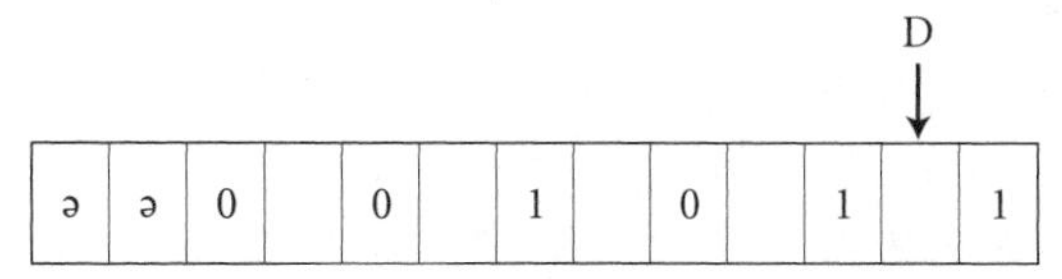

The machine, now in D, encounters a succession of blanks, which compel it to move ten squares to the left, where it encounters an ə. Its instruction upon encountering this symbol is to move one square to the right and shift into state E. The instructions for state E now require it to move, in total, twelve squares to the right, at which point it once again encounters a blank square. The machine now prints a 0, moves two spaces to the left, and reverts to state B.

													B↓	
ə	ə	0		0		1		0		1		1		0

Its next instruction, in state B, is to move one square to the right, print an x, and then move three squares to the left. This operation is repeated twice, producing two x's.

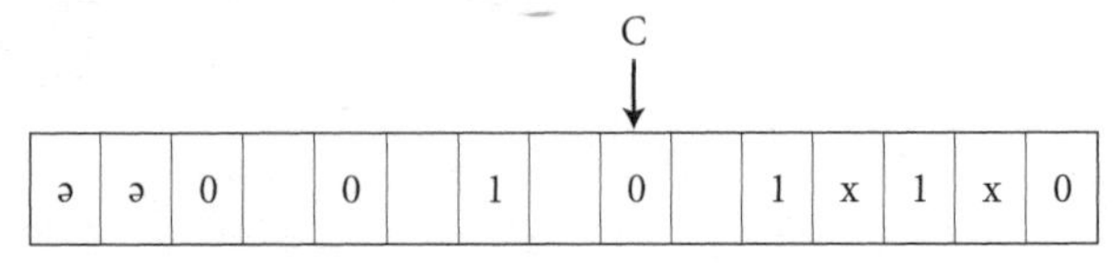

A series of further moves adds a 1, erases the x's, and adds another 1. The process repeats until we have:

ə	ə	0		0		1		0		1		1		0		1		1		1

We have now generated the beginning of the infinite sequence for which this particular machine is designed; we have 0010110111 . . . , written in alternate squares. As Turing notes, "the convention of writing the figures only on alternate squares is very useful: I shall always make use of it. I shall call the one sequence of alternate squares *F*-square and the other sequence *E*-squares. The symbols on the *E*-squares will be liable to erasure. The symbols on the *F*-squares form a continuous sequence." The *E*-squares provide the scratch paper on which the machine works out the basic operations that underlie the algorithm it is performing.

Read in unary notation, the sequence generated by Turing's *second* machine is simply the sequence of the natural numbers, with each number separated from the ones before and after it by a 0. The equation for which this machine provides an algorithm would thus be written

$$y = x + 1$$

with the machine endlessly putting in values for x and generating values for y.

Read in binary notation, on the other hand, we get the following sequence (with each number, again, separated by a 0 from the ones before and after):

$$1, 3, 7, 15, 31, 63, 127, \ldots$$

These numbers—each of which is written in binary notation as a sequence of 1's (1, 11, 111, 1111, 11111, 111111, etc.)—have in common that each is one less than a power of 2. The equation for which this Turing machine provides the algorithm can thus be written

$$y = 2^x - 1$$

with the machine, once again, calculating successive values of y.

It is worth noting that in both of Turing's examples, what the machine generates is not a computable number but a computable *sequence.* Most glosses on the Turing machine give examples such as the one with which I began, of machines that apply an algorithm to a particular set of values (in this case 2 and 2), find a solution (in this case 4), and then stop. Turing's machines, by contrast, go on forever, each one printing an infinite sequence of integers. Both types fit Stephen Kleene's definition of an algorithm as "a *finitely* described procedure, sufficient to guide us to the answer to any one of *infinitely* many questions, by *finitely* many steps in the case of each question." The first type, however, gives only one of the infinitely many answers, while the second gives all of them. Thus in the case of Turing's second example, each of the answers that the machine generates—1, 3, 7, 15, etc.—fits the definition of a computable number, while the numbers taken together fit the definition of an infinite computable sequence. Changing the configurations of the machine so that it answers for a specific n and then stops is a simple matter, and makes no substantive difference to Turing's thesis.

Comparatively speaking, both of these Turing machines are elementary. Indeed, one of Turing's principal points is that for every algorithmic procedure, no matter how complex, there exists a Turing machine the specific table of behavior for which will effect that algorithm. Each of these Turing machines would be defined by its table of behavior, the complexity of which would depend on the complexity of the algorithm in question. For certain algorithms, the table of behavior might require dozens of *m*-configurations and symbols. In his paper Turing sketches proofs for the algorithmic computability by Turing machines of values for π, *e*, all real algebraic numbers, and the real zeros of the Bessel functions.*

To simplify the process, Turing creates, in his paper, a sort of shorthand for writing out tables of behavior. He begins with the table for the first of his two machines, the one that prints out the sequence 0101010101:

***m*-Configuration**	**Symbol**	**Action**	**New *m*-Configuration**
A	blank	Print 0; move 1 square to the right	B
B	blank	Move 1 square to the right	C
C	blank	Print 1; move 1 square to the right	D
D	blank	Move 1 square to the right	A

*Turing computes π as $4[1- \frac{1}{3} + \frac{1}{5} - \frac{1}{7} + \frac{1}{9} \ldots]$. He computes *e*, an irrational number that plays an important role in, among other things, the calculation of natural logarithms, as $1 + 1 + \frac{1}{2!} + \frac{1}{3!} + \ldots$. As we have seen earlier, algebraic numbers are those irrational numbers that satisfy algebraic equations. The Bessel functions are the solutions to the Bessel differential equation: $x^2 \frac{d^2y}{dx^2} + x \frac{dy}{dx} 1 + (x^2 - m^2)y = 0$.

Turing now proposes giving numbers to the *m*-configurations, calling them q_1, q_2, q_3, q_4, . . . q_{r}. In addition, numbers are given to the symbols, which will be called S_1, S_2, S_3, S_4, . . . S_{r}. In particular, S_0 will signify a blank, S_1 will signify a 0, and S_2 will signify a 1. The table can now be rewritten as follows:

m-Configuration	Symbol	Action	New *m*-Configuration
q_1	S_0	P S_1, R	q_2
q_2	S_0	P S_0, R	q_3
q_3	S_0	P S_2, R	q_4
q_4	S_0	P S_0, R	q_1

Note that in this notation, "move right" is written as P S_0, R, meaning "print blank, then move right." A similar notation would take care of any E's (erasures).

The P's can now be removed, and the entire sequence rewritten on a single line:

$$q_1\ S_0\, S_1 R\ q_2;\ q_2\ S_0\ S_0 R\ q_3;\ q_3\ S_0\ S_2 R\ q_4;\ q_4\ S_0\ S_0 R\ q_1;$$

Turing next assigns to each symbol a letter according to the following scheme: q_i will be replaced by the letter D followed by *i* repetitions of the letter A, while S_j will be replaced by the letter D followed by *j* repetitions of the letter C. Right and left continue to be written as R and L, while "no move" is written as N. According to this system, $q_1\ S_0 S_1 R\ q_2$ would thus be expressed as *DADDCRDAA*, with *DA* replacing q_1, *D* replacing S_0, *DC* replacing S_1, and *DAA* replacing q_2. The sequence now reads

DADDCRDAA; DAADDRDAAA; DAAADCCRDAAAA; DAAAADDRDA;

Turing calls this the *standard description*, or S.D., of the machine. Yet he has one more transformation in mind. By assigning a numeral to each letter—1 for A, 2 for C, 3 for D, 4 for L, 5 for R, 6 for N, and 7 for ;—he is able to represent this standard description as a sequence of numerals.* The integer represented by these numerals he calls the *description number*, or D.N., of the machine. For the machine we have been discussing, the description number would be:

31332531173113353111731113322531111731111335317

This is obviously a very long number. Yet it is microscopic compared with the description numbers of more complex Turing machines. To make matters more difficult, this number would under ordinary circumstances be written in some version of binary notation, rendering it simpler for a machine to read but more arduous for a human being. None of which matters. Like the human computer whom it replaces, the Turing machine is in no hurry. On the contrary, because it lives in a hypothetical universe, untouched by human concerns such as speed or efficiency, it has all the time in the world.

*Turing's thesis, however, in no way depends upon the use of this particular coding system, and in fact other coding systems are far more economical. Nor is his coding system limited to these particular letters and numbers. For instance, one might assign the letter P (and the number 8) to the symbol –, indicating a negative number. One might also assign the letter Q (and the number 9) to the symbol /, indicating the dividing line between the numerator and the denominator of a fraction.

3.

Turing has now introduced and explained the idea of the *a*-machine and presented a system for encoding its instruction tables. He has also established that for every algorithmic procedure, an *a*-machine must, by definition, exist. And just as the sequence 001011011101111 . . . can be generated by the table we have given, "any computable sequence is capable of being described in terms of such a table." More importantly, "to each computable sequence there corresponds at least one description number, while to no description number does there correspond more than one computable sequence." Just as the computable sequences define the machines that generate them, each Turing machine defines a computable sequence. The uniqueness of the description numbers would even allow us to list them, as it were, alphabetically, starting at 0 and continuing on to infinity. On such a list, the machine for which we have just calculated, the description number would be in 313325311731133531117311133225311117311113353 17th place.

Does every machine, however, generate a valid computable sequence? The answer is no. Some machines, as Roger Penrose puts it, are "duds." An example of a dud machine proposed by Martin Davis would be one for which the instruction table reads roughly as follows:

m-Configuration	Symbol	Action	New *m*-Configuration
A	0	Move 1 square to the right; print 1	B
B	1	Move 1 square to the left; erase 0; print 1	C
C	1	Move 1 square to the right; erase 1; print 0	B

This machine would shuttle endlessly back and forth between 0 and 1; it would neither print a coherent sequence nor come to a stop. Such a machine Turing calls *circular*. On the other hand, a *circle-free machine* is one that is capable of generating a computable sequence. Apparently anticipating some head-scratching over the difference between a computable number and a computable sequence, Turing adds, "We shall avoid confusion by speaking more often of computable sequences than of computable numbers."

All the machines we have looked at, with the exception of the last, are circle-free machines. The first of these—the machine designed to add two numbers together—is fed with an input and then stops when it reaches its answer; the two machines that Turing gives as examples generate computable sequences. It would not, however, be difficult to design a variant on the machine that generates the sequence 001011011101111 . . . . Remember that this machine simply gives successive (and infinite) answers to the equation $y = 2^x - 1$ as the natural numbers from 1 onward are fed into it. The variant machine would be designed to plug one natural number at a time into the equation, halting when it gives each answer. Thus if one were to feed $x = 3$ into the machine, it would go through a process that would conclude with its generating the desired answer—7—and halt. But could one design a Turing machine that would analyze any *other* Turing machine and decide whether that machine was circular or circle-free? This question—known as the halting problem—lies at the heart of Turing's paper and leads directly into his analysis of the *Entscheidungsproblem*.

Turing's consideration of the halting problem brings him to what is unquestionably his most startling and original leap.

By way of proposing a method for investigating how one might determine whether a given Turing machine is circular or circle-free, he puts forward the idea of a "universal machine": a Turing machine that is capable of *imitating* the behavior of any *other* Turing machine, no matter what algorithm that machine is designed to perform. It is this hypothetical "universal machine" that really constitutes the prototype of the modern computer.

The section of Turing's paper that describes his universal machine begins with characteristic modesty. "It is possible," he writes, "to invent a single machine which can be used to compute any computable sequence. If this machine *U* is supplied with a tape on the beginning of which is written the S.D. of some computing machine *M*, then *U* will compute the same sequence as M." Remember that "S.D." stands for "standard description"—the sequence of letters into which the table of behavior for any given Turing machine can be translated, and which can in turn be translated into the integer (binary or denary) that is the machine's "description number."

By way of explaining the behavior of *U*, Turing begins by positing a third machine, M', "which will write down on the *F*-squares of the tape the successive complete configurations of *M*." Remember that earlier Turing defined a machine's "complete configuration" at any stage in its progress as comprising "the motion of the machine, the number of the scanned square, the complete sequence of all symbols on the tape, and the *m*-configuration." As an example of *M*, he recalls the second machine mentioned in the paper, the one that generates the sequence 001011011101111 . . . . Just as the table of behavior for this machine can be translated, with letters, into its standard description, so the sequence of moves that the machine goes through as it follows the rules prescribed by its table of behavior can be rewritten using letters. Once again, each *m*-configuration is written as *D* followed by the appropriate number of *A*'s, while each symbol is written as *D* followed by the appropriate number of *C*'s. As before, 0 is written as *DC*, 1 as *DCC*, and a blank as *D*. The symbol ə is written as *DCCC*, and though Turing does not mention the *x*, one can assume that it would be written as *DCCCC*.

Using this scheme, *M*' can print out the sequences of *F*-square symbols generated by *M*:

DA : *DCCCDCCCDAADCDDC* : *DCCCDCCCDAAAD-CDDC* : . . .

Turing's shorthand, here, requires a bit of unpacking. The first *DA* signifies that the machine begins in *m*-configuration A, which tells it to print the sequence əə00 and then shift into *m*-configuration B. The second sequence of letters describes the action taken as a result of this instruction, and amounts to an abbreviation of

əə B 0 blank 0

This is, of course, the sequence written on the *F*-squares at the point when the machine moves into *m*-configuration C, with the *m*-configuration that has generated this sequence (B) inserted between the two ə's and the actual figures that have been printed in the *F*-squares: two 0's and a blank. A colon separates this description from the next, which abbreviates

əəəə C 0 blank 0

This is the machine's state as it shifts into its next complete configuration. No new figures have been generated because, as we recall, *m*-configuration B, upon encountering a 0, does nothing; it simply tells the machine to shift into *m*-configuration C. In letters, this sequence reads

DCCCDCCCDAAADCDDC

The next complete configuration, following the implementation of *m*-configuration C upon encountering a 0, describes the machine having shifted two spaces to the right but remaining in C. This would look like

əə C 0 blank 0 blank blank

which would translate into

DCCCDCCCDAAADCDDCDDCDD

Now the machine, still in *m*-configuration C, finds itself at a blank square, at which point it is instructed to print a 1, shift one square to the left, and enter into *m*-configuration D:

əə D 0 blank 0 blank 1

Or:

DCCCDCCCDAAAADCDDCDDCC

Once again, colons separate descriptions of the complete configurations of the machine at each move that it makes.

A : əə B 0 blank 0 : əə C 0 blank 0 : əə C 0 blank 0 blank blank:
əə D 0 blank 0 blank 1

Or:

DA : DCCCDCCCDAADCDDC : DCCCDCCCDAAAD-CDDC : DCCCDCCCDAAADCDDCDD: DCCCDCCC-DAAAADCDDCDDCC

"It is not difficult to see," Turing concludes,

> that if *M* can be constructed, so can *M'*.The manner of operation of *M'* could be made to depend on having the rules of operation (*i.e.*, the S.D.) of *M* written somewhere within itself (*i.e.* within *M'*); each step could be carried out by referring to these rules. We have only to regard the rules as being capable of being taken out and exchanged for others and we have something very akin to the universal machine.

Turing here describes—again, with little fanfare—a prototype of the modern computer, in which the rules are stored "somewhere within" the machine, the software within the hardware, but can be "taken out and exchanged for others." It is a moment of insight made all the more extraordinary by its author's apparent failure to apprehend its implications.

Just one thing is lacking: "at present the machine *M'* prints no figures." It prints only representations (in the form of letters) of the figures that *M* would print. Turing corrects this omission by arranging for *M'* to print "between each successive pair of complete configurations the figures which appear in the new configuration but not in the old." The sequence thus becomes

DA: 0: 0: *DCCCDCCCDAADCDDC : DCCCDCCCDAAAD-CDDC* : . . .

Leaving out the two ə's, which, we recall, function only to indicate the initiation of the sequence, the only figures printed as a result of *m*-configuration B are a pair of 0's. At first it may seem slightly odd that these figures should be printed *before* the instruction that generates them—and yet we must remember that the machine we are dealing with is not *M*, but *M'*, and that the function of *M'* is not to generate the sequence but to *describe* the behavior of *M* as *it* generates the sequence. No figures appear between the second and the third descriptions because the third—*DCCCDCCCDAADCDDC*—does not, in fact, result in the generation of any more 0's or 1's. On the other hand, if we were to continue the sequence, we would soon see a 1:

DA : 0 : 0 : *DCCCDCCCDAADCDDC* : *DCCCDCCC-DAAADCDDC* : *DCCCDCCCDAAADCDDCDD* : 1 : *DCC-CDCCCDAAAADCDDCDDCC* : . . .

Our machine *M'* is now, in effect, operating as a universal machine, because in addition to the complete configurations of *M*, it is printing the computable sequence that *M* was designed to generate. With typical nonchalance, Turing concludes, "It is not altogether obvious that the *E*-squares leave enough room for the necessary 'rough work,' but this is, in fact, the case." Finally, he notes that "the sequences of letters between the colons . . . may be used as standard descriptions of the complete configurations. When the letters are replaced by figures . . . we shall have a numerical description of the complete configuration, which may be called its description number." For instance, the standard description of the complete configuration *DCC-CDCCCDAAAADCDDCDDCC* would translate into the description number 3222322231111323323322.

4.

Turing's next step is to set up the table of behavior for a universal machine, called *U*. At this point the symbolic language he employs—a combination of uppercase and lowercase German (Gothic) letters and lowercase Greek letters—becomes immensely confusing; indeed, to borrow a phrase from Roger Penrose, Turing's system of skeleton tables "would be rather more complicated to explain than the machine is complicated itself." I shall try in the following pages to offer a more reader-friendly précis of his ideas.

We recall that in order to employ the universal machine, *U*, we have to feed into it the description number of a specific Turing machine, *T*. The *m*-configurations listed on the table of behavior for *U* will now lead the machine through a series of maneuvers by means of which it is able to deduce from *T*'s description number the algorithmic process that *T* follows. *U* can then obtain the same result as *T*. Thus if *T* is the Turing machine that generates the sequence 0101010 . . . , and we want *U* to mimic *T*, we first feed into *U* the description number for *T*:

31332531173113353111731113322531111731111335317

U, in turn, acts upon that number according to the instructions specified by its own table of behavior, with the result that it generates the same sequence as *T*: 0101010 . . .

It is important to recognize here that within Turing's morass of German and Greek alphabets there lies a precise and detailed account of exactly how *U* functions: the sequence of *m*-configurations that it follows on its path to becoming, for a time at least, *T*. Moreover, there is no limit whatsoever to what *U* is capable of. Indeed, as Stephen Kleene notes, in subsequent years mathematicians have

> convinced themselves that all possible algorithms for computing number-theoretic functions can be embodied in Turing machines (Turing's thesis). The ingredients that are basically necessary were all provided by Turing: a fixed finite number of symbols, a fixed (perhaps very large) number of states, actions determined by the condition of one scanned square and the state in accordance with the constitution of the particular machine (i.e., its table), unlimited space (on the tape) for receiving the questions and reporting the answers and temporarily storing scratchwork, and of time (the moments) for completing the calculations.

In order to operate as a Turing machine, all that the universal machine requires is its description number. This means that in order to operate *U*, one would in principle need to draw up a list of the description numbers of all Turing machines so that these could be fed into *U* when required. (This list, of course, would include the description number for *U* itself—a description number as singular as any other. Using a coding somewhat different from Turing's, Roger Penrose provides the description number for *U*. It is 1,653 digits long and takes up an entire page of his book.) Some of these machines would be, by Turing's definition, circular, in that they would never produce computable numbers or computable sequences. But others would be circle-free. Turing defines the "number which is a description number of a circle-free machine" as a "*satisfactory* number." By analogy, the number that is the description number of a circular machine would be defined as an *unsatisfactory* number.

Now Turing is able to begin his attack on the *Entscheidungsproblem*. The question he asks is this: does there exist an algorithm (and hence a Turing machine) that can act upon the description number of *another* Turing machine in order to decide whether that number is satisfactory? In theory, such a machine (let's call it *D*) would be able to analyze the description number of a given Turing machine *M* and then draw a conclusion as to its viability. If it turned out that *M* was circular, *D* would end its computations by printing a 1. If it turned out that *M* was circle-free, *D* would conclude by printing a 0. Machine *D*, if it existed, would in some cases amount to a positive solution to the *Entscheidungsproblem*, in that its verdict on the circularity or noncircularity of some specific Turing machine would provide a judgment as to the decidability of the statement to which that Turing machine corresponded. For instance, a mathematician trying to determine the truth or falsity of Goldbach's conjecture would simply have to feed into it the description number of the machine designed to break even numbers down into primes, stopping only when it found one that did not split. If the machine printed a 1, he would know that Goldbach's conjecture was true. More generally, machine *D* could be used to test out *all* logical statements—even those that don't fit into the "try all numbers" model mentioned above—since by definition a proof is a series of axiom invocations and deductions based on rules of inference, and hence a mechanically checkable process. A proof must also consist of a finite numbers of sentences (or strings) each employing symbols from a finite symbolic alphabet. This means that, in theory, machine *D* could be fed with every possible combination of every possible symbol in the alphabet. It would take an extremely long time, but as we have noted, in Turing's world, time is not of the essence. *D* would simply check the possible strings, one by one, weeding out those that were nonsense and verifying those that were valid proofs.

But could such a machine exist? Turing addresses this question by taking the classic approach of reductio ad absurdum. That is to say, he begins with an assumption: let's say that *D* does, in fact, exist. We set *D* to work on the first Turing machine (M_0) in our list of all Turing machines, asking it to tell us whether, when fed with the whole number *m*, M_0 generates a computable sequence. In that case, *D* will print a 0. Otherwise *D* will print a 1. We then do the same with M_1, M_2, M_3, . . . and on to M_n, noting along the way which machines print 0 and which print 1. The machines that print 0 are the "good" machines, the ones that are circle-free.

Next we draw up a list of the computable sequences generated by the circle-free machines, from the first to the last. For each of these machines we write out the computable sequence that it generates as different whole numbers, starting with 0, are fed into it. This will obviously be a very long list; what is important is that it is exactly the sort of list that lends itself to a method invented by Georg Cantor during his investigations into infinity. This is known as the *diagonal method*.

Here's how Turing used it. Let us imagine that a chunk from somewhere in the middle of the list looks like this:

1	1	1	1	1	1	1	1	…
1	3	7	15	31	63	127	255	…
0	1	0	1	0	1	0	1	…
1	2	3	4	5	6	7	8	…
1	2	3	5	8	13	21	34	…
3	6	12	24	48	96	192	384	…
2	3	5	7	11	13	17	19	…
0	1	8	27	64	125	216	343	…

Bear in mind that this is a completely arbitrary list of actual computable sequences. The arrangement is also arbitrary, because it will have no bearing on what we are about to do.

We now generate a new sequence by cutting a diagonal swath through the diagram; that is to say, by taking the first number from the first sequence, the second number from the second sequence, etc.:

1	1	1	1	1	1	1	1	…
1	**3**	7	15	31	63	127	255	…
0	1	**0**	1	0	1	0	1	…
1	2	3	**4**	5	6	7	8	…
1	2	3	5	**8**	13	21	34	…
3	6	12	24	48	**96**	192	384	…
2	3	5	7	11	13	**17**	19	…
0	1	8	27	64	125	216	**343**	…

The new sequence we have generated is

1	3	0	4	8	96	17	343	...

We now add 1 to each of the numbers in this sequence. Our new sequence is then

2	4	1	5	9	97	18	344	...

Because the diagonal method is just the sort of algorithmic process for which one could design a Turing machine, this is obviously a computable sequence. Yet the list from which it was derived includes all the sequences that can be computed by circle-free Turing machines according to machine *D*—that is, all computable sequences—and that list cannot include our new sequence, because our sequence differs from the first sequence on the list in its first square, from the second in its second, etc. A contradiction has appeared. Therefore, there can be no machine *D*.

As Turing writes, this proof, "although perfectly sound, has the disadvantage that it may leave the reader with the feeling that 'there must be something wrong.'" After all, the number generated through the diagonal method can be described; then why can it not be computed? This is an important question, and one that he will address shortly. First, however, he offers an alternative proof that *D* cannot exist—one that has not the disadvantage of leaving the reader feeling as if she has fallen into a hole, and that also "gives a certain insight into the significance of the idea 'circle-free.'"

This alternate proof employs the universal machine. Let us imagine that we can somehow link the deciding machine, *D*, to the universal machine, *U*, thus creating a new hybrid machine, *DU*. Into this machine we feed the description number of arbitrary Turing machine *M*. *DU* now runs *D*, which determines whether *M* is circular or circle-free. If *M* is circular, the process stops, since there would be no point in feeding the description number of a circular machine into *U*, which would then replicate its circularity and run on forever. If, however, *M* turns out to be circle-free, *U* can be used to simulate its algorithmic action. Because *DU* includes this "checking" mechanism, by means of which it can make sure that *U* is fed only the description numbers of circle-free machines, *DU* itself is circle-free; that is to say, under no circumstance will it lapse into circularity.

Now we feed *DU*'s description number into *DU* itself. *D* quickly establishes that *DU*'s description number is, as Turing has shown, satisfactory—that is, that *DU* is circle-free. It therefore passes the description number onto *U*, which simulates the action of *DU*, feeding the description number of *DU* into *D*, which then passes it on to *U*, which then simulates the action of *DU* . . . and on and on. In other words, *DU*, when fed with its own description number, runs on forever. *DU* is circular. But we have just shown that *DU* is circle-free. And since it is impossible that *DU* be both, Turing writes, "we conclude that there can be no machine *D*." Decidability is impossible. We are back in the land of paradox, with Epimenides declaring that he is a liar and Bertrand Russell upsetting Frege's applecart.

5.

Turing's next maneuver in his journey toward a solution of the *Entscheidungsproblem* foreshadows a strategy he later employed in his work on artificial intelligence. He shows that if we can answer a simple question—does there exist a machine E which, when fed with the description number of arbitrary Turing machine M, will establish whether M ever prints a given symbol?—then we can also answer a more complex one: is there a machine that can establish whether a given logical formula is or is not provable? This step is necessary if Turing is to make his proof airtight, meeting the demands of mathematical rigor.

He begins, once again, by making a reductio ad absurdum assumption. Let's say that machine E exists, and that we want to use it to find out whether M ever prints a 0. We feed into E the description number of M, and it replies by telling us that M either does or does not print a 0 at some point during its action. To use Turing's own example, if the sequence that M prints out is

$$A\ B\ A\ 0\ 1\ A\ A\ B\ 0\ 0\ 1\ 0\ A\ B \ldots$$

then E will tell us that, yes, M does sometimes print 0's.

Next we construct a variant on M—M_1—which prints the same sequence as M but replaces the first 0 with another symbol; let's say a %. Thus where M prints the sequence

$$A\ B\ A\ 0\ 1\ A\ A\ B\ 0\ 0\ 1\ 0\ A\ B \ldots$$

M_1 prints the sequence

$$A\ B\ A\ \%\ 1\ A\ A\ B\ 0\ 0\ 1\ 0\ A\ B \ldots$$

Likewise we construct a machine, M_2, that replaces the first *two* 0's in the sequence printed by M with %'s:

$$A\ B\ A\ \%\ 1\ A\ A\ B\ \%\ 0\ 1\ 0\ A\ B \ldots$$

And on to $M_3, M_4 \ldots M_n \ldots$

We now construct another machine—*H*—which, when fed with the standard description of *M*, generates successively the standard descriptions of *M*, M_1, M_2, . . . M_n. (In a parenthesis Turing assures us that such a machine exists.) Combining *H* with our original "Is there a 0?" machine, *E*, we obtain a new machine, *HE*. When fed with the description number of *M*, *HE* first goes into its *H* mode and writes down *M*'s standard description. Shifting into *E* mode, *HE* then determines, from that standard description, whether *M* will ever print a 0. If the answer is that *M* never prints a 0, *HE* prints :0:. *HE* then goes through the same procedure for M_1, M_2, . . . M_n, in each case printing :0: if the machine is shown never to print a 0. The thing to remember is that *HE* will print :0: only in cases where *M* never prints any zeros (i.e., 1111111 . . .) or where *M* prints a finite number of 0's (e.g., 00011111111 . . .), in which case at some point in the iteration of M_1, M_2, . . . M_n, we will get a sequence along the lines of %%%1111111. . . . If, on the other hand, *M does* print an infinite number of 0's, *HE* will not print :0:.

The final step is reminiscent of Turing's earlier proof that machine *D* cannot exist: we feed into our hypothetical machine *E* the description number of *HE* itself. Remember that the function of *E* is to determine whether a given arbitrary Turing machine ever prints a given symbol, in this case 0. If *HE* itself is shown never to print a 0, then *M* must print 0 infinitely often. But if *HE* sometimes prints a 0, then *M* must print no 0's or a finite number of 0's. A similar process would allow us to determine whether *M* prints a finite or an infinite number of 1's. Because every computable number must contain an infinite number of 0's or an infinite number of 1's (or possibly both), Turing can now conclude, "By a combination of these processes we have a process for determining whether *M* prints an infinity of figures, i.e. we have a process for determining whether *M* is circle-free." Turing has already shown, however, that there can be no such process. Therefore machine *E* cannot exist.

It is now that Turing is able, at last, to establish the insolubility of the *Entscheidungsproblem.* Already he has demonstrated a method for expressing the action of a given Turing machine numerically. Now he explains how to represent that action in the form of a logical formula, which he calls Un (*M*). To the untrained eye, the symbolic language that he uses here appears, at the very least, daunting, yet the leap it represents is intuitively simple to grasp. He begins by showing how to describe simple aspects of *M* using logical statements and then coding those statements as logical formulae. For example, the statement "In the complete configuration *x* (of *M*) the symbol on the square *y* is *S*" would be coded as a logical formula that he calls $R_{S_1}(x,y)$. Likewise the statement "In the complete configuration *x* the square *y* is to be scanned" will be coded as $I\ (x,y)$, the statement "In the complete configuration *x* the *m*-configuration is q_m" will be coded as k_{q_m}, and the statement "*y* is the immediate successor of *x*" will be coded as $F(x,y)$. Using these subformulae, Turing can now write out the logical formula Un (*M*) and then show that Un (*M*) "has the interpretation 'in some complete configuration of *M*, S_1 (i.e., 0) appears on the tape." It follows that "if there is a general method for determining whether Un (*M*) is provable, then there is a general method for determining whether *M* ever prints a 0." Earlier Turing explained that in his usage, the expression "There is a general process for determining . . ." is equivalent to the expression "There is a machine which will determine . . ." We can therefore conclude that if there exists a machine to solve the *Entscheidungsproblem*, then there must also be a machine *E*. Indeed, this may be why Turing labeled this particular machine with the letter *E* in the first place. Since we know, however, that machine *E* cannot exist, we can conclude that the *Entscheidungsproblem* solver cannot exist either. Therefore, the *Entscheidungsproblem* cannot be solved. Using logic—and a miraculous machine—Turing has brought an end to the old ideal of a serene and traversable mathematical landscape.

4

God Is Slick

1.

Turing now had his result. He had provided a definition for a whole new category of numbers, the "computable numbers," and along the way proved the insolubility of the *Entscheidungsproblem*. More importantly, he had introduced into the discourse of mathematics a startling and original concept: the *a*-machine. "It is difficult to-day," Newman wrote in his memoir, "to realize how bold an innovation it was to introduce talk about paper tapes and patterns punched into them, into discussions of the foundations of mathematics. . . ." It was just as bold an innovation to talk about "states of mind" in a mathematics paper; as Hodges observes, the supplementary "instruction note" argument was in many ways a "safer" approach. Yet the problem of how human beings think had been on Turing's mind at least since 1931, when he wrote an essay entitled "Nature of Spirit" for Christopher Morcom's mother. The essay begins with a general account of the influence of developments in physics and quantum mechanics on the scientific conception of the universe, then moves quickly into the question of free will:

> We have a will which is able to determine the action of the atoms probably in a small portion of the brain, or possibly all over it. The rest of the body acts so as to amplify this. There is now the question which must be answered as to how the action of the other atoms of the universe are regulated. Probably by the same law and simply by the remote effects of spirit but since they have no amplifying apparatus they seem to be regulated by pure chance. The apparent non-predestination of physics is almost a combination of chances.

Chance, in other words, has in the age of quantum mechanics supplanted "spirit" as the guiding principle that must underlie any effort to understand the universe—or has it? Turing is clearly ambivalent on this point. Although the atoms, in their action, "*seem* to be regulated by pure chance" (italics mine), in fact they are "probably" subject to the same "will" by means of which we as human beings are able to control at least a small portion of our brains. Thus "the remote effects of spirit" have *not*, in fact, been banished.

What Turing appears to be struggling to reconcile here are his dedication to scientific rigor (a dedication partially instilled in him by Christopher Morcom) and his longing to preserve some link with Christopher's spirit after his death. Indeed, at this point the essay becomes much more personal, and though Christopher's name is never mentioned, his ghost hovers in the white spaces:

> Personally I think that spirit is really eternally connected with matter but certainly not always by the same kind of

> body. I did believe it possible for a spirit at death to go to a universe entirely separate from our own, but now I consider that matter and spirit are so connected that this would be a contradiction in terms. It is possible however but unlikely that such universes may exist.

Does the spirit survive the body? If so, how and where? The question is religious in nature, yet in discussing it, Turing makes a point of never lapsing into the language of mysticism, or sacrificing his objective "scientific" outlook. Certainly it would have been a comfort for him to imagine that Christopher Morcom's spirit, in some sense, had not just outlived his body but remained in the same "universe" as Turing:

> Then as regards the actual connection between spirit and body I consider that the body by reason of being a living body can "attract" and hold on to a "spirit," whilst the body is alive and awake the two are firmly connected. When the body is asleep I cannot guess what happens but when the body dies the "mechanism" of the body, holding the spirit is gone and the spirit finds a new body sooner or later perhaps immediately.
>
> As regards the question of why we have bodies at all; why we cannot live free as spirits and communicate as such, we probably could do so but there would be nothing whatever to do. The body provides something for the spirit to look after and use.

There is something intensely intimate and moving about this passage, written to comfort the mother of a boy whom Turing adored, and with whose "spirit" he hopes to remain "connected" by means of his *body*. More than two decades earlier, Forster had prefaced *Howards End* with the words "*Only connect. . . .*" The phrase reappears in chapter 22 of the novel, where Forster writes, "Only connect the prose and the passion, and both will be exalted, and human love will be seen at its highest. Live in fragments no longer." Forster's exhortation can be read as a call to connect the body and the spirit, and as such it puts him at odds with Clive Durham, who is able to sustain his relationship with Maurice only so long as sex plays no role in it; what scandalizes Clive, at novel's end, is not just the discovery that Maurice is about to run off with Clive's gamekeeper, Alec Scudder, but the announcement that Maurice and Alec have "shared." Turing, by contrast, concludes his essay with a passionate affirmation of the physical that suggests the possibility of Forster's having had an indirect influence on him. Were it not for the body, which "provides something for the spirit to look after and use," the spirit, presumably, will languish. Without the body to give it expression, the spirit remains an abstraction from which in the long run no sustenance can be derived.

As Hodges has noted, "Nature of Spirit" prefigures the investigations into the question of free will and determinism—the degree to which the spirit controls the body, and vice versa—that later provided the backbone for "Computable Numbers." After all, the two points of view with which Turing is concerned here exactly parallel the "state of mind" and "instruction note" arguments from "Computable Numbers." Yet the idea of investigating a "state of mind" might also have come from King's College. Indeed, John Maynard Keynes, in describing the "religion" of G. E. Moore's *Principia Ethica*, had also made use of the term "state of mind," writing that for him and his fellow Apostles, "states of mind were not associated with action or achievement or with consequences. They consisted in timeless, passionate states of contemplation and communion, largely unattached to 'before' and 'after.' " The value of these states of mind, Keynes continued,

> depended, in accordance with the principle of organic unity, on the state of affairs as a whole which could not be usefully analysed into parts. For example, the value of the state of mind of being in love did not depend merely on the nature of one's own emotions, but also on the worth of their object and on the reciprocity and nature of the object's emotions; but it did not depend, if I remember rightly, or did not depend much, on what happened, or how one felt about it. . . .

At first glance, Keynes's world of "passionate contemplation" and romantic friendship seems remote from Turing's, with its human computers performing algorithmic operations. Yet there is a common element. Both men were driven by an impulse to analyze the mental apprehension of experience, to break it down into discrete units: in one case "timeless, passionate states of contemplation and communion, largely unattached to 'before' and 'after,' " and in the other the "moments"—the *m*-configurations—into which a computational procedure can be subdivided. Also, under the surface of both analyses of the *mental*, there lies a tacit understanding that such states must take physical form if they are to have meaning. The body is implicit in Keynes's essay, just as the unspoken possibility that an *a*-machine might actually be built resonates in Turing's hypothetical, even abstract use of the machine analogy.

2.

Turing finished his first draft of "Computable Numbers" in the spring of 1936. It is hard to guess whether at this point he was aware of just how far-reaching its ramifications would be. In many ways he was as improbable a maverick as Gödel, from whose results he went to great pains to distinguish his own:

> It should probably be remarked that what I shall prove is quite different from the well-known results of Gödel. Gödel has shown that (in the formalism of Principia Mathematica) there are propositions **U** such that neither **U** nor –**U** is provable. As a consequence of this, it is shown that no proof of consistency of Principia Mathematica (or of **K**) can be given within that formalism. On the other hand, I shall show that there is no general method which tells whether a given formula **U** is provable in **K**, or, what comes to the same, whether the system consisting of **K** with –**U** adjoined as an extra axiom is consistent.

Turing's result emphasized process. True, his paper took as its starting point the Gödelian idea that mathematical operations involving numbers could be expressed *as* numbers. But then his fascination with the mind took Turing in a different direction from Gödel, a self-proclaimed "Platonist" and "mathematical realist" who once attempted an ontological proof of the existence of God. In many ways Gödel was an antiformalist. As he wrote in a letter to Hao Wang (December 7, 1967),

> I may add that my objectivistic conception of mathematics and metamathematics in general, and of transfinite reasoning in particular, was fundamental also to my other work in logic.
>
> How indeed could one think of *expressing* metamathematics *in* the mathematical systems themselves, if the latter are considered to consist of meaningless symbols which acquire some substitute of meaning only *through* metamathematics?

Unlike Turing's, Gödel's preoccupations were not of a sort that would inevitably lead him to the *Entscheidungsproblem.* Yet Turing's result, once he got there, had a distinctly Gödelian flavor: that is to say, his answer, in some sense, was no answer at all, since in effect all he had shown was that the decision problem itself was an example of an undecidable problem. On the other hand, Turing's paper was immensely constructive in that it set out a clear-cut theory of computability while giving specific examples of large classes of computable numbers.* The paper also put into circulation the first really usable model of a computing machine.† No

*As Casti and DePauli note, Turing lent support to Gödel's findings by showing "that Turing machines can calculate at most a *countable* set of numbers—that is, a set whose elements can be put into a one-to-one correspondence with a subset of the positive integers (the 'counting' numbers). But there are uncountably many real numbers; hence we come to the perhaps surprising result that the vast majority of real numbers are not computable."

†Hodges correctly urges caution in using the term "Turing machine," pointing out that the phrase "is analogous to 'the printed book' in referring to a class of potentially infinitely many examples. . . . Again, although we speak of 'the' universal Turing machine, there are infinitely many designs with this property."

matter that this machine, at least at first, was hypothetical; its simplicity was in many ways its greatest virtue.

In April 1936 Turing gave his draft of "Computable Numbers" to Newman. At first, according to Solomon Feferman, Newman was "skeptical of Turing's analysis, thinking that nothing so straightforward in its basic conception as the Turing machines could be used to answer this outstanding problem," but he soon came around and encouraged Turing to publish the paper. Turing was naturally elated. At twenty-four, he stood on the brink of making a major contribution to his discipline, the sort that would secure his position at Cambridge and lead to an increase on his rather paltry £300 per annum stipend. Everything seemed to be going swimmingly. And then a difficulty arose.

That May, Newman received by post an offprint of an article by the Princeton mathematician Alonzo Church entitled "An Unsolvable Problem of Elementary Number Theory." The paper introduced a system called the lambda calculus, developed by Church in conjunction with his students Stephen Kleene and John Barkley Rosser, and then used that system to propose a definition of "λ-definability" that was in effect synonymous with Turing's of computability. Worse for Turing, in a second paper Church used the concept of λ-definability to show that the *Entscheidungsproblem* was insoluble. The first paper, though it had been presented before the American Mathematical Society on April 19, 1935, had taken a year to cross the pond. The second had appeared in the *Journal of Symbolic Logic* just as Turing was finishing the first draft of "Computable Numbers."

Newman shared the news of Church's papers with Turing, for whom it came as a shock. Once again—as at Sherborne, as with Sierpinski, as with his dissertation—history had pipped him at the post. Church, he explained to his mother, was "doing the same things in a different way." But did this mean that his own paper was unpublishable? Newman, to Turing's great relief, thought not. On the contrary, he told his mother, "Mr. Newman and I have decided that the method is sufficiently different [from Church's] to warrant publication of my paper too." Newman even suggested that Turing go to Princeton in order to study with Church, and toward that end wrote Church a letter outlining the situation:

> An offprint which you kindly sent me recently of your paper in which you define "calculable numbers," and shew that the Entscheidungsproblem for Hilbert logic is insoluble, had a rather painful interest for a young man, A. M. Turing, here, who was just about to send in for publication a paper in which he had used a definition of "Computable Numbers" for the same purpose. His treatment—which consists in describing a machine which will grind out any computable sequence—is rather different from yours, but seems to be of great merit, and I think it of great importance that he should come and work with you next year if that is at all possible.

Newman was worried, even at this early stage in the game, that Turing's habit of working in isolation might end up hurting his career. Afterward, in his memoir, Newman wrote that his former student's "strong preference for working everything out from first principles instead of borrowing from others . . . gave freshness and independence to his work, but also undoubtedly slowed him down, and later on made him a difficult author to read." As a rule, most mathematicians labor alone, their only tools a pencil and a pad (or a blackboard and a piece of chalk); even so, in mathematical circles, an excess of self-imposed isolation tends to be frowned upon. Working on his own, as Turing did, involved a trade-off. On the one hand, as Gandy would later argue, "it is almost true to say that Turing succeeded in his analysis because he was not familiar with the work of others. . . . The approach is novel, the style refreshing in its directness and simplicity. . . . Let us praise the uncluttered mind." On the other, Turing's ignorance of what his contemporaries were up to meant that Church's paper took him by surprise. As it happened, concern with the question of computability was very much in the mathematical air in the mid-1930s. Not only Church, Kleene, and Rosser but also Gödel, Jacques Herbrand, and Emil Post had been at work on the problem, which each described by means of his own terminology: Herbrand's "effective calculability" was equivalent to Church's "λ-definability," Gödel's concept of the "recursive function,"* and Turing's of the computable number, just as Post's formulation for a "finite-1 process" (worked out in cognizance of Church's work, though not of Turing's, and also published in the *Journal of Symbolic Logic* in 1936) bore a startling resemblance to Turing's *a*-machine:

*Jacques Herbrand (1908–1931) coined the term "effectively calculable" just before his death, in a skiing accident, at the age of twenty-three. According to Church, the work that he himself conducted in conjunction with Kleene and Rosser had its origin in the lectures on the general recursive function that Gödel gave at Princeton in 1934, for which Kleene and Rosser took the notes. Casti and DePauli give a good definition of the recursive function, calling it "a function for which there is a mechanical rule for computing the values of the function from previous values, one after the other, starting with some initial value."

> In the following formulation . . . two concepts are involved: that of a *symbol space* in which the work leading from problem to answer is to be carried out, and a fixed unalterable *set of directions* which will both direct operations in the symbol space and determine the order in which those directions are to be applied.

Rather than using the metaphor of the machine, Post envisioned a sort of factory divided up into "boxes" in which "the problem solver or worker is to move and work . . . being capable of being in, and operating in but one box at a time. And apart from the presence of the worker, a box is to admit of but two possible conditions, i.e., being empty or unmarked, and having a single mark in it, say a vertical stroke." Although Post was American and taught at City College in New York, his frame of reference was the same mass-manufacturing ethos that informs *The Man in the White Suit.** Yet his sequence of boxes is literally analogous to Turing's tape, just as his "worker" is analogous to the *a*-machine. Indeed, the exactitude of the parallel between Post's formulation and Turing's machine lends credence to the Platonic conception of mathematics as a process of discovery rather than invention. It was as if an idea were issuing forth from nature itself, avid to find expression. And though Church had the very real advantage of being the first out of the starting gate, it was not yet clear whether his lambda calculus would, in the end, prove to be

*Post, who had lost an arm in an accident as a child, was probably manic-depressive. He died in 1954, at the age of fifty-seven, after suffering a heart attack during an electroshock therapy session. He took a highly unusual approach to mathematics, emphasizing psychology and intuition.

the most usable, the most pragmatic, or the most compelling of the approaches in circulation.

3.

Church himself was a decidedly odd figure. Born in Washington, D.C., in 1903, he had spent virtually his entire adult life at Princeton, earning his bachelor's degree, master's degree, and Ph.D. from the university before joining the faculty in 1929. The only period of time he spent away from Princeton was in 1927 and 1928, when, as a National Research Fellow, he studied at Harvard, in Göttingen with Hilbert, and in Amsterdam with Brouwer. In a memoir of his years at Princeton, the Italian mathematician Gian-Carlo Rota recalls Church as looking "like a cross between a panda and a large owl. He spoke slowly in complete paragraphs which seemed to have been read out of a book, evenly and slowly enunciated, as by a talking machine. When interrupted, he would pause for an uncomfortably long period to recover the thread of the argument."

Church was famous for working all night, leaving his notes—carefully marked with colored pencils—for the mathematics department secretary to find and type up in the morning. He contributed little to the department aside from teaching and editing the reviews section of the *Journal of Symbolic Logic*, which he helped found in 1936; indeed, his failure to show up for faculty meetings often raised eyebrows and may have been part of the reason why he was not promoted to the rank of full professor until 1947—eighteen years after he joined the faculty.

Stories circulated about Church's remoteness. His colleague Albert Tucker recalled being told by the dean of the faculty at Princeton "that he often met Church crossing the campus, and he would speak to Church and Church would not speak in reply." When afternoon tea was served in the departmental common room, Church would arrive "toward the end of the tea session, and he would take any milk or cream that was left in the pitchers there and dump that into one of the almost-empty teapots and drink this mixture of milk and tea. Then he would depart for his office, where he would work through the night." His lecture style was pedantic and painstaking to a fault, leading to the witticism "If Church said it's obvious, then everybody saw it half an hour ago."*

His behavior tended toward the compulsive. For example, Rota recalls, he had a vast collection of science-fiction novels, each of which was marked cryptically with either a circle or a cross. In some cases he had made corrections to wrong numberings in the margins of the table of contents. His lectures, Rota continues, invariably "began with a ten-minute ceremony of erasing the blackboard until it was absolutely spotless. We tried to save him the effort by erasing the board before his arrival, but to no avail. The ritual could not be disposed of; often it required water, soap, and brush, and was followed by another ten minutes of total silence while the blackboard was drying." The lectures themselves required nothing in the way of preparation, since they consisted of literal recitations of typewritten texts prepared over the course of twenty years and kept in Fine Hall library. On those rare occasions when Church felt obliged to diverge from the prepared text, he would warn his students in advance.

*The joke—taking in other prominent Princeton mathematicians—continues, "If Weyl says it's obvious, von Neumann can prove it. If Lefschetz says it's obvious, it's false."

To some degree, Church's punctiliousness was part and parcel of his talent as a logician; as Tucker notes, "he was completely oblivious of everything that went on in the world except in mathematical logic." According to Rota, Church would never make such a simple statement as "It is raining," because "such a statement, taken in isolation, makes no sense. . . . He would say instead: 'I must postpone my departure for Nassau Street, inasmuch as it is raining, a fact I can verify by looking out the window.'" The lambda calculus was similarly flawless in its precision; in Kleene's words, it boasted "the remarkable feature that it [was] all contained in a very simple and almost inevitable formulation, arising in a natural connection with no prethought of the result." Still, Kleene argued, "for rendering the identification with effective calculability the most plausible—indeed, I believe compelling, Turing computability [had] the advantage of aiming directly at the goal. . . ." Church's "λ-definitions" might, as Turing himself modestly put it, be more "convenient," but the *a*-machine was "possibly more convincing."

Oddly enough, the smaller the academic arena, the higher the stakes tend to be. In 1936, the discipline of mathematical logic not only had few adherents but was in somewhat bad odor in the larger mathematical community, particularly in America. As Church recalled in an interview with William Aspray conducted in 1984, "there were not many others interested in this field, and it was thought of as not a respectable field, with some justice. There was a lot of nonsense published under this heading." That few mathematicians would take much interest in something as arcane as the *Entscheidungsproblem*, however, did not make its resolution any less significant for the two men each of whom saw himself, justly, as the victor in a battle that had been going on for centuries. Moreover, both needed the recognition. In 1936, after all, Church was only thirty-three years old, just nine years older than Turing. He was still an assistant professor and had no other means of support for himself and his family (including his aged father, a retired judge) beyond his Princeton salary. In mathematics, advancement comes about only by way of significant publications, vetted by authorities after careful review. Being able to lay claim to the *Entscheidungsproblem* by means of "Church's thesis" mattered as much to Church as being able to do so by means of "Turing's thesis" meant to Turing.* Yet in the interview with Aspray in 1984, when Church was eighty, he was curiously evasive on the question of how he first heard about Turing and "Computable Numbers." Indeed, the exchange is as telling for what it leaves out as for what it includes:

> *Aspray*: If you don't mind, I would like to ask a few more questions about this topic, because it is one of particular interest to me since I wrote my dissertation on Turing. How did you hear about Turing's work?
>
> *Church*: Well, Turing heard about mine by seeing the published paper in the *American Journal of Mathematics*. At the time his own work was substantially ready for publication. It may already have been ready for publication. At any rate he arranged with a British periodical to get it published rapidly, and about six months later his paper appeared. At the same time, I think, Newman in England wrote to me about it.

*Interestingly, both names remain in circulation, though the thesis is more commonly known today as the "Church-Turing thesis."

Aside from errors in chronology, what is striking about this passage is that Church answers the question "How did you hear about Turing's work?" as if it were the question "How did *Turing* hear about *your* work?" For Turing, Church tells Aspray, the news of his own results came "as a great disappointment." We never learn how Church took the news of Turing's results.

One point was beyond dispute: although the methods that Church and Turing employed could not have been more different—indeed, it was the uniqueness of Turing's methods that made his paper so striking—the conclusions they reached were identical. This meant that Turing would have to acknowledge Church before he could publish "Computable Numbers," and so that August he sketched out, as an appendix to the paper, a proof of equivalency between his notion of computability and Church's of λ-definability. He then sent the manuscript off. Church appeared willing to have him at Princeton, and on September 23 Turing's mother saw him off at Southampton, where he boarded the Cunard liner *Berengaria*, traveling steerage. Among the items that he brought with him were an old violin picked up at the Farringdon Road market in London and an antique sextant. "Of all the ungainly things to hold," his mother wrote, "commend to me an old-fashioned sextant case. Though some readings were taken, what with the movement of the ship, and a defect in the instrument and Alan's inexperience, he doubted their accuracy." Turing headed a letter sent from on board the *Berengaria* 41° 20' N. 62° W.

He arrived in New York on September 29. Since the establishment, in 1932, of the Institute for Advanced Study,

Princeton had rapidly become the Göttingen of the twentieth century, and though the institute remained a separate entity from the university's mathematics department, the fact that both were housed at Fine Hall rendered the distinction academic. Turing wrote to his mother,

> The mathematics department here comes fully up to expectations. There is a great number of the most distinguished mathematicians here: J. v. Neumann, Weyl, Courant, Hardy, Einstein, Lefschetz, as well as a host of smaller fry. Unfortunately there are not nearly so many logic people here as last year. Church is of course, but Gödel, Kleene, Rosser and Bernays who were here last year have left. I don't think I mind very much missing any of these except Gödel. Kleene and Rosser are, I imagine, just disciples of Church and have not much to offer that I could not get from Church. Bernays is I think getting rather "vieux jeu" that is the impression I get from his writing, but if I were to meet him I might get a different impression.

Bernays, in fact, had been one of Hilbert's disciples at Göttingen; as recently as 1930 he had voiced his optimistic faith that a positive solution to the *Entscheidungsproblem* would eventually be found. As for Hardy, as a fellow Cambridge homosexual, he would have made, one might think, a likely mentor for Turing. Instead, Turing reported, "he was very standoffish or possibly shy. I met him in Maurice Pryce's* rooms the day I arrived, and he didn't say a word to me. But he is getting much more friendly now."

Like most of the graduate students in the mathematics

*Maurice Pryce (1913–2003), physicist and professor at Oxford.

One of a set of passport photos taken of Alan Turing in the 1930s. (King's College Library, Cambridge)

department, Turing spent nearly all his time in Fine Hall, a three-story red-brick building with elaborate casement windows and a slate roof. Notwithstanding its gothic fripperies, Fine Hall had been open only since 1931. The mathematician Oswald Veblen (1880–1960), Church's mentor and the guiding spirit behind its construction, as well as the nephew of the economist Thorstein Veblen, intended for it to emulate the architecture of Oxford and Cambridge. Although he was from Iowa and of Norwegian descent, Veblen had distinct Anglophilic tendencies; he had taught at Oxford and was married to Elizabeth Richardson, the sister of the British physicist Owen Willans Richardson. Perhaps for this reason, he conceived of

Fine Hall as a sort of Oxbridge college in its own right, albeit one intended exclusively for the use of mathematicians and physicists. Thus, just as at Oxford colleges there were junior common rooms in which the students could mingle with the faculty, and senior common rooms in which the dons could gather alone to drink port, at Fine Hall there was a common room (analogous to the junior common room) open to everybody (it was situated so that one had to pass through it on the way to the library), as well as a room reserved for the exclusive use of the professors on the principle "not always understood by those who try to bring about closer relations between faculty and students that in all forms of social intercourse the provisions for privacy are as important as those for proximity." In this "professors' room," faculty members could chat or read in front of an elaborate wooden fireplace the surround of which was carved with mathematical imagery, including a fly exploring the Moebius strip; over the mantel was inscribed a quotation from Einstein—"Raffiniert ist der Herr Gott, aber boshaft ist er nicht"—which the mathematical physicist H. P. Robertson translated as "God is slick, but he ain't mean." Some of the windows were divided into polyhedra, while in others the leaded glass was etched with formulae, including $E = mc^2$.

As much effort was made to mimic Oxbridge ritual as Oxbridge architecture; thus at the Graduate College, students wore gowns to dinner, while in the Fine Hall common room, tea was served every afternoon at three. And yet in spite of the tea, the common room had a distinctly casual (and distinctly American) tone that was entirely unlike that at King's College. Graduate students with little money spent nearly all their time there, returning to their furnished rooms only to sleep. There were no tea ladies or waiters; instead, the teas were organized and served by fellowship students, in compensation for the fact that they earned more money and had less work to do. Off the room was a kitchenette with an electric stove and—wonder of wonders—a dishwasher; here there was kept a large quantity of "cookies" (for Turing a foreign term) ordered in bulk from the National Biscuit Company, which would soon become Nabisco. Particularly during the depression, there had to be a quota on the number of cookies served, in order to discourage hungry graduate students from making a meal of them.

Not surprisingly, few professors actually used the room reserved for their exclusive occupancy. The common room was much more fun. Games were always being played there—go, chess, and kriegspiel (a variety of blindfold chess) as well as games invented on the premises, such as psychology, a card game of which Turing's friend Shaun Wylie was immensely fond. Wylie, who hailed from Oxford (his father had been a "greats" don), was finishing the last of his three years at Princeton just as Turing was starting the first of what would turn out to be his two. Rather quickly, Wylie drew him into his circle of English and American graduate students, the other members of which included Francis Price, Will Jones, and Bobby Burrell. The group organized treasure hunts, play readings, even an (ad-) hockey team, which played matches against the girls at Miss Fine's School (Miss Fine was the sister of Dean Fine, for whom Fine Hall was named) and Vassar. But Turing, though he was looked upon with affection, and considered an "honorary member of the clique," remained a bit aloof. "I suspect that he was glad to be involved," Wylie later told Frederick Nebeker, "but was certainly at that stage not a leading spirit."

All told, Princeton proved rather flummoxing for a young middle-class Englishman like Turing. He was puzzled, for example, that none of the graduate students with whom he worked minded "talking shop. It is very different from Cambridge in that way." American speech habits also took him aback:

> These Americans have various peculiarities in conversation which catch the ear somehow. Whenever you thank them for anything, they say "You're welcome." I rather liked it at first, thinking I was welcome, but now I find it comes back like a ball thrown against a wall, and become positively apprehensive. Another habit they have is to make the sound described by authors as "Aha." They use it when they have no suitable reply to a remark, but think that silence could be rude.

The more fluid aspects of American social interaction were particularly disconcerting for a young man raised in a society defined by class distinctions, physical reticence, and rigid notions of propriety. "Though prepared to find democracy in full flower," Mrs. Turing wrote, "the familiarity of the tradespeople surprised [Turing]; he cited as an extreme case the laundry vanman who, while explaining what he would do in response to some request of Alan's, put his arm along Alan's shoulder. 'It would be just incredible in England.'" Of course, reading between the lines, one wonders if behind Turing's dismay at the vanman's "familiarity" there lay a more profound unease: surprise at encountering frank erotic possibility, uncertainty as to how to respond to it, even retrospective disappointment at an opportunity lost.

What news of Mother England he received came through American newspapers. "I am sending you some cuttings about Mrs. Simpson," he wrote to his mother on November 22, "as representative sample of what we get over here on this subject. I don't suppose you have ever heard of her, but some days it has been 'front page stuff' here." A little more than a week later he was complaining that he was "horrified at the way people are trying to interfere with the King's marriage. It may be that the King should not marry Mrs. Simpson, but it is his private concern. I should tolerate no interference by bishops myself and I don't see that the King need either."

Strong words, especially to a mother. Yet Turing had reason to identify with the plight of Edward VIII; his love stories, after all, were also of a sort likely to draw disapproval from bishops. The hypocrisy of the Church of England outraged him, as did the apparent effort of the British press to suppress the story. "I believe the government wanted to get rid of him and found Mrs. Simpson a good opportunity," he wrote. Clearly the idea that political institutions might use a man's personal life against him neither shocked nor particularly surprised Turing, though it disturbed him, as did what he called the "disgraceful" behavior of the archbishop of Canterbury: "He waited until Edward was safely out of the way and then unloaded a whole lot of quite uncalled-for abuse. He didn't dare do it whilst Edward was King. Further he had no objections to the King having Mrs. Simpson as a mistress, but marry her, that wouldn't do at all."

It was just the sort of sexual hypocrisy that Turing had grown up with, at schools where homosexual behavior was tolerated so long as it was never named, or allowed to evolve into homosexual identity. And how swiftly quiet tolerance turned into brutal suppression, once the lovers decided to go public! The archbishop's "Don't ask, don't tell" policy galled Turing not just on behalf of the king and Mrs. Simpson but because of the implications it held for his own future. Indeed, as late as May of 1939, he was still fretting over the abdication, noting to his mother that he was "glad that the Royal Family are resisting the cabinet in their attempt to keep Edward VIII's marriage quiet."

At Princeton, meanwhile, there were parties to go to—the sort at which the fate of Mrs. Simpson might well be the topic of conversation. Although mathematicians don't usually have a reputation for being party people, the crowd at Fine Hall was famously sociable. In particular, the spirited and glamorous John von Neumann and his second wife, Klara, gave grand parties at their house to which the graduate students were often invited. Hermann Weyl and his wife, Hella, held gatherings at which Turkish coffee was served. By contrast, the occasional dinners that Church hosted with his wife, Mary, were somewhat dreary affairs, at least to judge from what Turing reported to his mother. "Church had me out to dinner the other night," he wrote to her that October. "Considering that the guests were all university people I found the conversation rather disappointing. They seem, from what I can remember of it, to have discussed nothing but the different states they came from."

As for Church, if he took any notice at all of Turing, he later forgot it. Many years later, William Aspray asked him to name the graduate students with whom he worked in the thirties. The answer he gave is noteworthy, once again, for the one name it omits: Turing's. Reminded of Turing by Aspray, Church continued, "Yes, I forgot about him when I was speaking about my own graduate students. Truth is, he was not really mine. He came to Princeton as a grad student and wrote his dissertation there." When asked to describe Turing's personality, Church said, "I did not have enough contact with him to know. He had the reputation of being a loner and rather odd." The same, of course, had often been said of Church himself.

4.

"Computable Numbers" was published by the *Proceedings of the London Mathematical Society* in January 1937. To Turing's disappointment, the response was decidedly underwhelming. "I have had two letters asking for reprints," he wrote to his mother, one from his old Cambridge friend R. B. Braithwaite

> and one from a proffessor [*sic*] in Germany. . . . They seem very much interested in the paper. I think possibly it is making a certain amount of impression. I was disappointed by its reception here. I expected Weyl who had done some work connected quite closely with it some years ago, at least to have made a few remarks.

But Weyl, whose 1918 monograph *Das Kontinuum* had been a landmark text in classical analysis, said nothing. Neither, apparently, did the dashing and cosmopolitan John von Neumann, like Weyl, a former disciple of the Hilbert program. Von Neumann was in attendance at the 1930 address in Königsberg at which Gödel had announced his incompleteness theorem, and after the talk had approached Gödel asking for details; according to Solomon Feferman, he "was one of the first to appreciate the significance of Gödel's incompleteness results. In fact, it is reported that he obtained the second incompleteness theorem . . . independently of Gödel, once he had learned of Gödel's first incompleteness theorem."

Von Neumann was famous not just for his astounding mathematical aptitude but for the catholicity of his interests—in a discipline notable for its compartmentalization, he was something of a jack-of-all-trades—and when Gödel's findings were published, he abandoned logic altogether in favor of other fields, once going so far as to claim never to have read another paper in the subject after 1931. Most of his colleagues doubted that this was true; in any case, his apparent allergy to logic may have been behind his failure to respond, either positively or negatively, to "Computable Numbers."

Ironically, part of the problem was that in 1937 Princeton was such an important center for mathematics and physics. More and more the Institute for Advanced Study was becoming an escape route for European scientists forced to flee their homelands in the wake of Nazism. As a result, the community's prestige increased in direct proportion to the drain of émigré faculty from such former European powerhouses as Göttingen. Rarely had so many great minds been gathered in a single building. As Joseph Daly explained to Aspray, "You had von Neumann, you had Einstein, you had Veblen, you had Knebelman and T. Y. Thomas and Al Tucker. Everything was going on at once, and the real problem for graduate students was to keep from getting so diverted into 16 different fields that you didn't get anything done."

Logicians were already in the minority—a state of affairs that Gödel's absence and von Neumann's decampment only worsened. Nor did solving the *Entscheidungsproblem* look like quite such a big deal when you had Albert Einstein just down the hall. In order to get noticed at Princeton, you had to have done something, but you also had to know how to promote yourself, and in this arena Turing—shy from the outset—was far less adept than his friend Maurice Pryce, of whom he had written to his mother, "Maurice is much more conscious of what are the right things to do to help his career. He makes great social efforts with the mathematical big-wigs." That Turing had come second in the succession after Church, who was on the Princeton faculty, only added to his difficulty, as did the fact that his paper made for such slow reading. As Newman had pointed out, his isolation, though it lent freshness to his thinking, also made his prose impenetrable. All told, it took time to dig deep enough inside "Computable Numbers" to recognize the startling originality at its heart, and few of his colleagues had the patience.

And the paper really was something new. When they were first performed, Chopin's compositions perplexed and even outraged his listeners. Monet's paintings horrified the Parisians of his epoch. "Computable Numbers" elicited no reactions so violent, but it was ahead of its time, and in a sense the silence that greeted it reflected the inability of the age in which Turing was working to register the implications of what he had done. In 1936, after all, Post's factory analogy represented a far more recognizable symbolic language than Turing's *a*-machine. As for Church, though his lambda calculus boasted a certain elegance, as well as the virtue of self-containment, it displayed none of the boldness, the imaginative vigor, that made Turing's formulation so compelling and so memorable.

For his part, Church was more than generous to his putative student. Reviewing "Computable Numbers" in the *Journal of Symbolic Logic*, he wrote,

> As a matter of fact, there is involved here the equivalence of three different notions: computability by a Turing machine, general recursiveness in the sense of Herbrand-Gödel-Kleene, and λ-definability in the sense of Kleene and the present reviewer. Of these, the first has the advantage of making the identification with effectiveness in the ordinary (not explicitly defined) sense evident immediately—i.e. without the necessity of proving preliminary theorems. The second and third have the advantage of suitability for embodiment in a system of symbolic logic.

Notably, Church's review also marked the first use of the term "Turing machine"—evidence even this early of the degree to which the machine, above and beyond the problem it had been invented to solve, was taking on a life of its own. Such a machine was appealing to visualize, tirelessly shifting the tape back and forth between its jaws. To put the matter in theatrical terms, the machine was starting to steal the show from all the other characters—from Hilbert, from Church, even from its inventor.

In the meantime, Turing worked under Church on a paper intended to lay out clearly the equivalence of Turing computability and λ-definability. It was published in the fall of 1937 in the *Journal of Symbolic Logic*. In the circle of those who cared about such things, some discussion went on as to which system would in the end prove to be the most useful. Kleene, for one, felt that it was "actually easier to work with general recursive functions than with complicated Turing-machine tables," while Post criticized some of the details of Turing's paper. None of his colleagues went so far as to deny the importance of Turing's work. And yet none of them embraced it, either.

It was clear that at least for the time being—and at least at Princeton—the Turing machine would have to take a back seat to the other approaches. The kingdom might be tiny, but Church ruled it, and Turing knew enough not to contemplate regicide. Instead, he worked away on his Ph.D. thesis, which Church directed, and wrote two papers on group theory, one of them in response to a problem passed on by von Neumann. (It seemed that von Neumann was perfectly willing to give Turing encouragement so long as the work in question had nothing to do with logic.) He also gave a lecture on computability before the Mathematics Club, but the attendance was small. "One should have a reputation if one hopes to be listened to," he wrote to his mother. "The week following my lecture G. D. Birkhoff* came down. He has a very good reputation and the room was packed. But his lecture wasn't up to the standard at all. In fact everyone was just laughing about it afterwards." It seemed that Turing was rapidly learning a painful lesson: too often, reputation trumps talent. What remained to be seen was whether he would develop the schmoozing skills already practiced by his friend Maurice Pryce.

5.

Social awkwardness aside, Turing's abilities did not go unnoticed at Princeton. In particular he received encouragement

*Although George David Birkhoff (1884–1944) was considered one of the greatest mathematicians of his time, he was also, in Einstein's words, "one of the world's greatest anti-Semites," consistently keeping Jews out of his department at Harvard.

from Luther Eisenhart, dean of the Department of Mathematics, and his wife, Katherine. As he wrote to his mother in February 1937,

> I went to the Eisenharts regular Sunday tea yesterday and they took me in relays to try and persuade me to stay another year. Mrs. Eisenhart mostly put forward social or semi-moral semi-sociological reasons why it would be a good thing to have a second year. The Dean weighed in with hints that the Procter Fellowship was mine for the asking (this is worth $2000 p.a.) I said I thought King's would probably prefer that I return, but gave some vague promise that I would sound them on the matter.

He did sound King's out on the matter—to no avail. The college failed to come through with a lectureship (Maurice Pryce won one instead), and in the end—rather reluctantly, since he was becoming homesick—Turing agreed to stay the extra year. Von Neumann wrote him a letter of recommendation for the Procter—notably, he emphasized Turing's work in the theories of almost periodic functions and continuous groups, but said nothing about "Computable Numbers"—and this cinched the deal. As Turing wrote to his mother, he would now be "a rich man." He nonetheless decided to return to England at least for the summer, and on June 23—his twenty-fifth birthday—he sailed out of New York.

Back in Cambridge, Turing kept himself busy correcting some errors in "Computable Numbers," working on his Ph.D. thesis, and completing the paper he was writing under Church establishing the equivalency of his own notion of computability, λ-definability, and Gödel's concept (taken up by Kleene) of "general recursiveness." Perhaps the most interesting work he undertook, however, involved a mathematical problem that was distinctly remote from the rarefied world of logic. This was the Riemann hypothesis, at that time an issue of pressing concern among number theorists. As of this writing, the Riemann hypothesis remains unsolved; indeed, it is considered the most important unsolved problem in mathematics.

The Riemann hypothesis concerns the distribution of the prime numbers. Thanks to Euclid, we know that there are infinitely many primes. But is there a pattern to the sequence in which they crop up? At first the distribution appears to be arbitrary:

1, **2**, **3**, 4, **5**, 6, **7**, 8, 9, 10, **11**, 12, **13**, 14, 15, 16, **17**, 18, **19**, 20, 21, 22, **23**, 24, 25

Moreover, as you climb up the number line, there seem to be fewer and fewer primes:

N	**Number of primes from 1 to** *N*	**% of primes from 1 to** *N*
10	4	40
100	25	25
1000	168	16.8
10000	1,229	12.29
100000	9,592	9.59

In fact, though, there *is* some pattern to the primes, as Karl Gauss (1777–1855) discovered around 1793, when he was fifteen. Gauss recognized a correlation between the distribution of the primes up to a certain number n and the natural logarithm of that number. As a result he was able to come up with the *prime number theorem*, a formula for determining the *number* of primes up to a certain number—or something close to it: inevitably the formula overestimated by a small degree the number of primes. The next step was to find a way of eliminating the "error term" in the prime number theorem—that is to say, to find a formula by means of which the *exact* number of primes up to even some inconceivably huge *n* could be calculated.

This was where the German mathematician Bernhard Riemann (1826–1866) came in. Riemann was working with the so-called zeta function, denoted by the Greek letter ζ and calculated according to the formula

$$\zeta(x) = \frac{1}{1} + \frac{1}{2} + \frac{1}{3^x} \ldots \frac{1}{\text{nx}} \ldots$$

By feeding complex numbers* into the zeta function, Riemann discovered, he could eliminate the error term in the prime number theorem. Through some pretty difficult mathematics, he was able to hypothesize that whenever the zeta function, upon being fed with a complex x, took the value of 0, the real part of the complex x would be $\frac{1}{2}$. This meant that on a graph of the function, all the zeros would be positioned along the so-called *critical line* of $\frac{1}{2}$. The elaborate mathematical link that Riemann mapped between the zeros of the zeta function and the prime numbers meant that if his hypothesis

*A complex number is defined as the combination of a real number and a so-called *imaginary* number. An imaginary number is, quite simply, the square root of a negative number—imaginary because any number, positive or negative, when multiplied by itself, gives a positive result. Hence $\sqrt{-1}$ cannot exist, and it is "imaginary." $\sqrt{-1}$—the basis of all the imaginary numbers—is referred to as i. By the same token, $\sqrt{-4}$ is $2i$, $\sqrt{-9}$ is $3i$, etc. Because they combine real with imaginary numbers, complex numbers are expressed as $3 + 6i$, $-2.547 - 1.34i$, etc.

was true, the error term in the prime number theorem could be eliminated.

But was the Riemann hypothesis true? Could it be proven? Disproof would be simple, a matter of finding just one zero off the critical line. Unfortunately, testing out the zeros of the Riemann hypothesis was a technically challenging business, requiring the use of highly complex mathematics. A formula developed by G. H. Hardy and J. E. Littlewood, for instance, allowed them to confirm by the late 1920s that the first 138 zeros lay on the critical line, but turned out to be infeasible for the calculation of zeros beyond that number. Likewise a project undertaken in 1935 by the Oxford mathematician E. C. "Ted" Titchmarsh (1899–1963) to map the zeros using punch-card machines intended for the calculation of celestial motions established only that the first 1,041 zeroes lay along the critical line. Only one contrary example would have been needed to disprove the hypothesis. A proof, on the other hand, would have to be theoretical, eliminating entirely the possibility of there being even a single zero off the line, up to infinity. (Of course, a third possibility was that the hypothesis would turn out to be true yet unprovable in a Gödelian sense. So far, however, no one had been able to prove its unprovability, either.) That Riemann himself was rumored to have had a proof of the hypothesis—burned, along with other important papers, by an overzealous housekeeper after his death—only added to the sense of urgency and frustration surrounding the problem. Opinions on the validity of the hypothesis varied wildly; even the opinion of Hardy, who devoted much of his career to wrestling with Riemann, fluctuated at various points in his career. In 1937, the year he was living in Princeton, he was apparently feeling some pessimism as to the truth of the hypothesis—a pessimism he may well have passed on to Turing.

Turing's work that summer involved a particularly arcane aspect of the Riemann hypothesis. Although application of Gauss's prime number theorem suggested that the theorem would always *over*estimate the number of primes up to *n*, in 1933 the Cambridge mathematician Stanley Skewes had shown that at some point after $10^{10^{1034}}$ a crossover occurred, and the formula began to *under*estimate the number of primes. Hardy made the observation that $10^{10^{1034}}$ was probably the largest number ever to serve any "definite purpose" in mathematics, and tried to indicate its enormousness by noting,

> The number of protons in the universe is about 10^{80}. The number of possible games of chess is much larger, perhaps $10^{10^{50}}$. If the universe were the chessboard, the protons the chessmen, and any interchange in the position of two protons a move, then the number of possible games would be something like the Skewes number.

Turing's ambition was either to lower the bound established by Skewes or to disprove the Riemann hypothesis altogether. But though he wrote up a draft of a paper, he never published his results. After his death the paper circulated, and a number of errors were discovered, all of which were corrected by A. M. Cohen and M. J. E. Mayhew in 1968. Using Turing's methods, they were able to lower the bound established by Skewes to $10^{10^{529.7}}$. However, it then turned out that in 1966 R. S. Lehman had reduced the bound to 1.65×10^{1165} by another method. Even in death, it seemed, Turing was fated to be beaten to the punch.

6.

When Turing returned to Princeton in the fall of 1937, Hardy was no longer in residence; he had gone back to Cambridge, where he was beginning work on his memoir, *A Mathematician's Apology*, which would be published in 1940. All friendliness aside, he and Turing had never really gotten on, as in Hodges' words, "a generation and multiple layers of reserve" divided them. Hardy was also a far more orthodox mathematician than Turing. His work fell very much within the purview of pure mathematics, which was why he could argue, in *A Mathematician's Apology*, that

> the "real" mathematics of the "real" mathematicians, the mathematics of Fermat and Euler and Gauss and Abel and Riemann, is almost wholly "useless" (and this is as true of "applied" as of "pure" mathematics). It is not possible to justify the life of any genuine professional mathematician on the ground of the "utility" of his work.

Hardy saw mathematics as fundamentally innocent, even neutral—a sort of Switzerland of the sciences—and this meant that as late as 1940 he was still taking "comfort" in the knowledge that "no one has yet discovered any warlike purpose to be served by the theory of numbers or relativity, and it seems very unlikely that anyone will do so for many years." As a pure mathematician, he wanted to affirm his faith that his discipline would remain forever "gentle and clean," even as in Germany the Third Reich was adapting a typewriter-sized encipherment machine called the Enigma—in production since 1913—for military use, and in America the theory of relativity was being applied to the production of an atomic bomb. The simple construction of an apparently unbreakable code, and a machine to transmit it, would be the first of many tasks for which mathematicians would find themselves drafted in the war effort; far from being allowed to pursue their "gentle and clean" science, they would now have to put their skills at the disposal of rival military machines, the one bent on world domination and the other on stopping it. Indeed, Hardy was so utterly wrong in his affirmation that one cannot help wondering how much of what he wrote was wishful thinking. For soon the fate of Europe would be resting, to a great extent, on the shoulders of a small group of scientists. Their decisions would save lives and cost lives. Alan Turing was to be one of them.

For most of his twenty-five years, he had been trying to reconcile his passion for mathematical logic with his impulse to build things. As a homosexual, he was used to leading a double life; now the closeted engineer in him began to clamor for attention, and not long after arriving at Princeton, he wrote to his mother,

> You have often asked me about possible applications of various branches of mathematics. I have just discovered a possible application of the kind of thing I am working on at present. It answers the question "What is the most general kind of code or cipher possible," and at the same time (rather naturally) enables me to construct a lot of particular and interesting codes. One of them is pretty well impossible to decode without the key, and very quick to encode. I expect I could sell them to H. M. Government for quite an

additional sum, but I am rather doubtful about the morality of such things. What do you think?

Questions of morality aside, the project intrigued Turing so much that once he was settled again in Princeton, he undertook a plan to build, at the machine shop overseen by the physics department, an electronic multiplier employing relay-operated switches specifically for the purpose of encipherment. The use of such switches, in 1937, was in and of itself something of a novelty; so was the use of binary numbers, which in turn allowed for the employment of a two-valued Boolean algebra, with TRUE and FALSE represented by 0 and 1 and 0 and 1, in turn, associated with the on and off positions of the switches. In this regard Turing was replicating the research of a contemporary, Claude Shannon (1916–2001), whose 1937 MIT master's thesis, *A Symbolic Analysis of Relay and Switching Circuits*, would be the first published work to describe the implementation of Boolean algebra in a switch-based machine.

As Turing explained to the physicist Malcolm MacPhail, the idea behind the encipherment machine was to "multiply the number corresponding to a specific message by a horrendously long but secret number and transmit the product. The length of the secret number was to be determined by the requirement that it should take 100 Germans working eight hours a day on desk calculators 100 years to discover the secret factor by routine search!" According to MacPhail, Turing actually built the first three or four stages of this project. He didn't know, of course, that in just a few years, the Second World War would erupt, and he would be sent by his government to Bletchley Park, where he would undertake the design of a much more sophisticated *de*cipherment machine. Already, however, he was on the road to what would become the preoccupying focus of the next dozen years for him: the application of mathematical logic to the building of machines.

Although the electronic multiplier, which Turing never finished, was his principle spare time project for the academic year 1937–38, he had not forgotten the Riemann hypothesis. Titchmarsh's punch-card machine had succeeded in establishing only that the first 1,041 zeroes of the zeta function were on the critical line of $\frac{1}{2}$. Now Turing proposed building a machine based on one in Liverpool that carried out the calculations necessary to predict tidal motions. He had recently learned about a powerful method for calculating the zeros that Riemann himself had come up with, and that the mathematician Carl Ludwig Siegel (1896–1981) had rediscovered among the papers that Riemann's wife, Elise, had managed to rescue from his pyromaniacal housekeeper. Like Titchmarsh before him, Turing had noticed a parallel between Riemann's formula and those that were applied to the prediction of tides, planetary orbits, and other such physical phenomena. The Liverpool machine was an analog machine—it functioned by replicating the tidal motions it was intended to calculate—and in principle a similar machine for the calculation of the zeros of the zeta function could likewise run on . . . either forever or until it came up with a zero off the critical line. Despite Titchmarsh's written encouragement, however, Turing appears not to have had the chance to put his ideas to the test—at least while he was at Princeton.

Officially, he was devoting himself to the thesis he was writing for Church. This was to be a treatise on a problem relating to Gödel's incompleteness theorem. In the wake of

Gödel's result, much effort was being expended in logic circles on finding ways to minimize its impact, so that it should interfere as little as possible with the practice of mathematics. Such a project, of course, was very much within the purview of Alonzo Church, and Turing undertook it very much at Church's suggestion. Church was at first not entirely convinced by the incompleteness theorem, and spent several years in pursuit of the mathematical equivalent of loopholes. What really galled him about Gödel, however, was that Gödel appeared simply not to think very highly of the lambda calculus; indeed, at one point Gödel went so far as to tell Church that he "regarded as thoroughly unsatisfactory" Church's "proposal to use λ-definability as a definition of effective calculability." By way of reply, Church suggested that if Gödel could "present any definition of calculability which seemed even partially satisfactory," then he (Church) "would undertake to prove that it was included in the lambda-calculus." Which, no doubt, he could have: the point Church was missing was that Gödel's objections to the lambda calculus were philosophical, even aesthetic in nature, and crystallized as soon as he read "Computable Numbers."

Gödel had not met Turing, and never would; in 1937 he was living in the Josefstadt, the doctors' district of Vienna, and would return to Princeton only in the fall of 1938, by which point Turing had gone back to England. Nonetheless, he was soon making it clear how much he preferred Turing's formulation to Church's.* The universality and directness of the *a*-machine—the fact that, in Kleene's words, it "aimed

*According to Kleene, "only after Turing's formulation appeared did Gödel accept Church's thesis, which had then become the Church-Turing thesis."

directly at the goal"—appealed to the Platonist in Gödel, who in 1946 applauded Turing for

> giving an absolute definition of an interesting epistemological notion, i.e. one not depending on the formalism chosen. In all other cases treated previously, such as demonstrability or definability, one has been able to define them only relative to a given language, and for each individual language it is clear that the one thus obtained is not the one looked for.

By "definability," Gödel was no doubt alluding to Church's λ-definability, which he considered so "unsatisfactory." On the other hand, Gödel later wrote that "due to A. M. Turing's work, a precise and unquestionably adequate definition of the general concept of formal system can now be given. . . ." As Feferman points out, Gödel never voiced publicly his dissatisfaction with Church's thesis. Even so, he made it known how he felt. For Church, receiving a failing grade from the greatest figure in his field must have stung.

Whether Gödel's obvious preference for Turing's approach inclined Church to keep his distance from his ostensible student remains a matter of speculation. In certain ways they were very much alike—both eccentric, solitary, slightly out of step with ordinary social intercourse. Whereas Church, however, was genuinely antisocial—indeed, almost Aspergerian in his rigidity and disconnection—Turing voiced without shame a hunger for friendship and love, for Forsterian connection, that only his sexual anxiety stymied. He was also deeply pragmatic; he wanted to understand how daisies grew, and to invent patent inks, and to build typewriters and computers.

Church, by contrast, lived almost entirely in his own head. Daisies and typewriters were to him what horned sheep and white sheep were to Boole: figurines in the game of logic. Earlier in the century, Bertrand Russell had asserted that "pure mathematics is the subject in which we do not know what we are talking about, or whether what we are saying is true." Church subscribed to the formalist tradition, insisting that mathematical symbols be rigorously emptied of any semantic content that might inhere in them. Such a perspective was deeply troubling to Gödel, given that he made a point of "distanc[ing] himself from . . . formal-syntactic interpretations of science and mathematics." In the end, it is no less surprising that Turing should have found working with Church a limiting and, finally, frustrating experience than that Gödel—the self-proclaimed realist—should have chafed at the systematic exclusion of semantic "meaning" that was in Church's own view the lambda calculus's great virtue.

5
The Tender Peel

1.

It was at this point that Alan Turing's intellectual life began to diverge from the course on which it appeared set. Having turned down an offer to work as John von Neumann's assistant at Princeton (at a salary of $1,500 a year), he returned in fall 1938 to England, where he was recruited to join a course on cryptography and encipherment sponsored by the Government Code and Cipher School in London. Somehow word of his interest in codes and code breaking—not to mention his talent at mathematics—had found its way to Commander Alastair Denniston, the school's director. Although far from a rabid patriot, Turing had no qualms about lending his services to the government; war appeared likely, and he was gravely distrustful of Hitler. The decision to take the course marked the beginning of a long association with the British government that would, in Hodges' words, have "fateful" repercussions, in that it would require Turing, for the first time, to surrender "a part of his mind, with a promise to keep the government's secrets."

Nineteen thirty-eight also saw the British premiere of the Disney version of *Snow White and the Seven Dwarfs*—a film with which, curiously enough, both Turing and Gödel became fascinated. Turing went to see *Snow White* with his friend David Champernowne and took an especially keen pleasure in the scene where the Wicked Queen immerses the apple in her poisonous brew. "Dip the apple in the brew, / Let the sleeping death seep through," she chants, then cackles to her sidekick, a raven, as the poison forms a skull on the surface of the apple. "Look at the skin," the Queen continues.

A symbol of what lies within.
Now turn red to tempt Snow White,
To make her hunger for a bite.

The Queen offers the apple to the raven, who flaps wildly, trying to escape. "It's not for you, it's for Snow White," the Queen says, laughing.

When she breaks the tender peel,
To taste the apple from my hand,
Her breath will still, her blood congeal,
Then I'll be the fairest in the land!

The scene captivated Turing to such a degree that he took to chanting the Queen's verses over and over. Perhaps what appealed to him was its morbid eroticism, not to mention the rather blatant allusion to the biblical myth: this apple tempts Snow White as another tempted Eve. Yet whereas Eve bites into the apple of her own free will (she could, after all, have resisted the serpent's blandishments), Snow White is the victim of a well-rehearsed campaign of deception on the part of the Queen, who dresses as a hag, and tries out a variety of ploys to persuade Snow White to taste the apple; indeed, she is in the end able to convince her nemesis only by telling her that the apple is "magic" and that it will grant Snow White her wish. Only by appealing to her passion for the Prince can the Queen break down Snow White's resistance and persuade her to break "the tender peel," in so doing losing her psychic virginity and consigning herself to a "sleeping death." Which aspect of the film's fraught psychosexual architecture appealed to Turing, however, he never let on.

Over the Christmas break, Turing attended another training course sponsored by the Government Code and Cipher School, after which he visited the school every two to three weeks to assist in the work being done there. But he had not forgotten his Princeton project to build a machine to calculate the zeros of the Riemann zeta function, and from Cambridge he applied for a grant from the Royal Society to pay for the machine's construction, naming Hardy and Titchmarsh as references. On the application form, he wrote,

> Apparatus would be of little permanent value. It could be added to for the purpose of carrying out similar calculations for a wider range of *t* and might be used for some other investigations connected with the zeta-function. I cannot think of any applications that would not be connected with the zeta-function.

The emphasis Turing put on the machine's lack of "permanent value" underscored its remoteness from the universal machine. That machine, after all, could perform *any* algorithm presented to it, whereas the machine Turing wanted the money to build was so single-purpose as to defy any effort to think up further uses for it. So desultory, in fact, is Turing's statement that one wonders if perhaps he hoped his proposal would be turned down. For better or worse, though, the Royal Society granted him £40, and he settled down in earnest to constructing the device by means of which he hoped to prove, once and for all, the falsity of the Riemann hypothesis.

His initial plan was to mimic the design of the Liverpool machine, which used a system of strings and pulleys to simulate the periodic sine waves corresponding to the motions of the tides and then added up their values; an answer could be obtained by measuring the lengths of string as they wrapped around the pulleys. But after consulting with Donald MacPhail, a student of mechanical engineering at Cambridge and the brother of Turing's Princeton friend Malcolm MacPhail, he changed his mind and decided that instead of strings and pulleys, he would replicate the circular motions of the zeta function by means of gear wheels that meshed together. Like the Liverpool machine, this would be an analog machine, replicating motions in order to measure them. By contrast, digital machines (of which the early computers are examples) work by manipulating symbols and can therefore be put to much more general use. That the machine Turing envisioned was analog, by definition limited its applicability.

The money was there, though, so MacPhail drafted a blueprint (now in the Turing archive at King's College), and for a time the floors of Turing's rooms were strewn with precision-cut gear wheels awaiting their eventual incorporation into the machine. But the project was destined, once again, to be left unfinished. The war intervened. In fact, it was not until 1950 that Alan Turing was finally able to use a machine to test the zeros of the Riemann zeta function. This one would be digital.

2.

The spring semester of 1939 at Cambridge, Turing was involved in two courses, both entitled "The Foundations of Mathematics." The first, an investigation into the history of mathematical logic, he taught himself. It had fourteen students, though Turing told his mother that he suspected attendance would drop off as the term advanced. The second, an investigation into the philosophical basis of mathematics, was taught by the Austrian philosopher Ludwig Wittgenstein. The participants included, in addition to Turing, several people with remarkable names: R. G. Bosanquet, Yorick Smythies, Rush Rhees, Marya Lutman-Kokoszynska, and John Wisdom. Notes were taken by Bosanquet, Rhees, Smythies, and Norman Malcolm, an American graduate student in philosophy who later wrote a remarkable memoir of the philosopher.

Wittgenstein was an eccentric figure. The scion of a wealthy as well as intellectually gifted family (his father was an engineer and something of a titan in the Austrian steel and iron industry), he counted Russell and Frege among his intellectual mentors and was close friends with Keynes and Hardy. Because he had been educated at Trinity College, Cambridge, he spoke flawless English, and in 1937 had assumed the chair in philosophy at Cambridge previously held by G. E. Moore. He had inherited a fortune upon the death of his father in 1912, but had given away most of his money, much of it in the form of a large anonymous grant for the promotion of literature. He also acted as something of a benefactor for the poet Rainer Maria Rilke.

Like Turing, Wittgenstein was a "confirmed solitary," frequently retreating to a farmhouse in rural Norway where he could write and think in seclusion. His desire to flout the values of the world in which he had been raised led him, for a time, to work as a schoolmaster. During the First World War he joined the Austrian army as a volunteer, eventually being captured by the Italians, who held him as a prisoner of war for the better part of a year. The manuscript of one of his seminal works, the *Tractatus Logico-Philosophicus*, was in his rucksack at the time. Fortunately he was able to send copies to Russell and Frege, thanks chiefly to the intervention of John Maynard Keynes.

In 1939 Wittgenstein was fifty. Malcolm, who got to know him around this time, writes that when he first met the author of the *Tractatus Logico-Philosophicus*, he expected him

> to be an elderly man, whereas this man looked *young*—perhaps about thirty-five. His face was lean and brown, his profile was aquiline and strikingly beautiful, his head was covered with a curly mass of brown hair. I observed the respectful attention that everyone in the room paid to him. . . . His look was concentrated, he made striking gestures with his hands as if he were discoursing. All the others maintained an intent and expectant silence. I witnessed this phenomenon countless times thereafter and came to regard it as entirely natural.

In sharp contrast to Alonzo Church's lectures, which never varied from the prepared text in the library at Fine Hall,

Wittgenstein's "were given without preparation and without notes." He told Malcolm "that once he had tried to lecture from notes but was disgusted with the result; the thoughts that came out were 'stale,' or, as he put it to another friend, the words looked like 'corpses' when he began to read them." Though he spoke "with the accent of an educated Englishman," his dress was astonishingly informal, given the time and the place. "He always wore light grey flannel trousers, a flannel shirt open at the throat, a woolen lumber jacket or a leather jacket. . . . One could not imagine Wittgenstein in a suit, necktie or hat."

Wittgenstein taught his class in a highly unconventional manner. For one thing, the course was long, thirty-one hours divided into twice weekly sessions over the span of two terms. Generally speaking, the meetings took place in his own rooms in Whewell's Court, Trinity College, with the students sitting on the floor or bringing in chairs. The rooms

> were austerely furnished. There was no easy chair or reading lamp. There were no ornaments, paintings, or photographs. The walls were bare. In his living-room were two canvas chairs and a plain wooden chair, and in his bedroom a canvas cot. . . . There was a metal safe in which he kept his manuscripts, and a card table on which he did his writing. The rooms were always scrupulously clean.

Certain rules governed attendance. Although Wittgenstein placed no limitations on who could participate, prospective students had to submit first to an interview with the philosopher over tea, during which there would be long, intimidating silences, as Wittgenstein loathed small talk. Students also had to commit to attending the course continuously, not just for one or two meetings. Late arrivals were frowned on. "One had to be brave to enter after the lecture had begun," Malcolm recalls, "and some would go away rather than face Wittgenstein's glare." Though Wittgenstein could be hard on his students—once telling Yorick Smythies, "I might as well talk to the stove!"—he could be equally hard on himself, punctuating his lectures with exclamations such as "I'm a fool," "You have a dreadful teacher," or "I'm just stupid today." According to Malcolm, "lecture" was hardly the right term to describe the classes, since they consisted principally of exchanges between Wittgenstein and the participants, dialogues broken only by the prolonged silences to which those in attendance soon became acclimated.

The subject of the course was the relationship of mathematics to what Wittgenstein called "ordinary" language and, by extension, "ordinary" life. As John Casti and Werner DePauli explain, "Wittgenstein applied his own principle that 'the boundaries of my language signify the boundaries of my world' to mathematics. In other words, the objects of mathematics were understood as being limited to those entities that could be formulated in mathematical language." The approach was made clear at the first lecture, in which—perhaps because Turing was the lone mathematician among the participants—Wittgenstein used him as an example:

> Suppose I say to Turing, "This is the Greek letter sigma," pointing to the sign σ. Then when I say, "Show me a Greek sigma in this book," he cuts out the sign I showed him and puts it in the book.—Actually these things don't happen. These misunderstandings only immensely rarely arise—

> although my words might have been taken either way. This is because we have all been trained from childhood to use such phrases as "This is the letter so-and-so" in one way rather than another.
>
> When I said to Turing, "This is the Greek sigma," did he get the wrong picture? No, he got the right picture. But he didn't understand the application.

One of Wittgenstein's ambitions was to compel his students to recognize the importance of common sense even in philosophical inquiry. ("Don't treat your common sense like an umbrella," he told them. "When you come into a room to philosophize, don't leave it outside but bring it in with you.") Nor was it accidental that of all the participants in the seminar, it was Turing he singled out, time and again, to serve as the representative of what might be called the logicist position; Wittgenstein was, in his own words, always trying to "tempt" Turing toward making claims that favored logic over common sense (though not always with success). As a practicing mathematician, Turing could be counted on to reiterate the traditional postulates of his discipline and in so doing give Wittgenstein the opportunity to pull the rug out from under them. Church, or someone like him, would have made a more convenient whipping boy, and had Wittgenstein known more about the unorthodoxy of some of Turing's ideas, he might have taken a different tack. Instead, he assigned Turing to the role of the naysayer, as this bit of dialogue from the sixth lecture shows:

> *Turing* [asked whether he understood]: I understand but I don't agree that it is simply a question of giving new meanings to words.

Wittgenstein: Turing doesn't object to anything I say. He agrees with every word. He objects to the idea he thinks underlies it. He thinks we're undermining mathematics, introducing Bolshevism into mathematics. But not at all.

We are not despising the mathematicians; we are only drawing a most important distinction—between discovering something and inventing something. But mathematicians make most important discoveries.

This was just one of many instances, during the lectures, when Wittgenstein took on, dismantled, and in a certain sense recontextualized the arguments underlying pure mathematics, including the old debate as to whether mathematics was invention or discovery.* Though he was far from being an antirealist or an intuitionist of the Brouwer school, he insisted that his students question even the most fundamental axioms of arithmetic. ("We say of a proof that it convinces us of a logical law.—But of course a proof starts somewhere. And the point is: What convinces us of the primitive propositions on which the proof is based? Here there is no proof.") In

*Hardy, though not a participant in the seminar, was often cited as a sort of avatar of traditional mathematical thinking; Wittgenstein took particular interest in the distinction that Hardy drew between "knowing" and "believing in" a mathematical theorem or hypothesis, such as Goldbach's conjecture. He liked to turn Hardy's ideas about proof upside down, as in this passage from the fourteenth lecture: "Professor Hardy says, 'Goldbach's theorem is either true or false.'—We simply say the road hasn't been built yet. At present you have the right to say either; you have a right to *postulate* that it's true or that it's false.—If you look at it this way, the whole idea of mathematics as the physics of the mathematical entities breaks down. For which road you build is not determined by the physics of mathematical entities but by totally different considerations."

Wittgenstein's rooms, the most basic assumptions were subject to scrupulous analysis:

> What is counting? Pointing to things and saying "1, 2, 3, 4"? But I need not say the numbers: I could point and say "Mary had a little lamb" or I might whistle "God Save the King" or anything.—But normally the process of counting is used in a different way, whereas "Mary had . . ." is not used in this way at all. If you came from Mars you wouldn't know.

By the same token, for someone from Mars, an axiom of mathematics might be "that whenever numerals of more than five figures cropped up . . . they were thrown away and disregarded." Or that $5^{(6^4)}$ is the same as $(5^6)^4$. The point was to interrogate the relationship between the mathematical meaning of words and their "ordinary" meaning, and to expose the consequences of divorcing one from the other. For example, in the third lecture, he imagined a wallpaper factory in which the pattern on the paper consisted of the proof that $21 \times 36 = 756$ repeated over and over. "You might call this figure the proof that $21 \times 36 = 756$," Wittgenstein told his students,

> and you might refuse to recognize any other proof. Why do we call this figure a proof?
>
> Suppose I train the apprentices of wallpaper manufacturers so that they can produce perfect proofs of the most complicated theorems in higher mathematics, in fact so that if I say to one of them "Prove so-and-so" (where so-and-so is a mathematical proposition), he can always do it. And suppose that they are so unintelligent that they cannot

make the simplest practical calculations. They can't figure out if one plum costs so-and-so, how much do six plums cost, or what change you should get from a shilling for a twopenny bar of chocolate.—Would you say that they had learnt mathematics or not?

They know all the calculations but not their applications. So one might say, "They have been taught pure mathematics."

In short, though the apprentices "would use the words 'proof,' 'equals,' 'more,' etc., in connexion with their wallpaper designs, . . . it would never be clear why they used them." On the other hand, "if it were said, 'The proof of Lewy's guilt is that he was at the scene of the crime with a pistol in his hand'—what is the connexion between this and calling the [wallpaper] pattern a proof? They wouldn't know why it was called a proof."

Wallpaper hangers, soldiers and generals, white lions, collapsing bridges: Wittgenstein was always presenting analogies in his lectures, asking his students to "suppose" one thing or another. While Wittgenstein could jump easily from one analogy to the next, however, Turing tended to cling tenaciously to the examples as if, by virtue of their simple iteration, they had taken on a kind of physical reality for him. In the transcripts of the course, no other student responds to Wittgenstein so frequently, or so readily, as Turing does, and in many cases his responses amount to proposing extensions of the analogy with which Wittgenstein has begun. On the example of the wallpaper hangers, for instance, Turing says, "The ordinary meanings of words like 'three' will come out to some extent if they are able to do simple things like counting the numbers of symbols in a line." Likewise the examples that Turing himself gives suggest the degree to which he was becoming more and more preoccupied with the relationship of logic to events in the real world:

> *Turing:* One could make this comparison between an experiment in physics and a mathematical calculation: in the one case you say to a man, "Put these weights in the scale pan in such-and-such a way, and see which way the lever swings," and in the other case you say, "Take these figures, look up in such-and-such tables, etc., and see what the result is."
>
> *Wittgenstein:* Yes, the two do seem very similar. But what is this similarity?
>
> *Turing:* In both cases one wants to see what will happen in the end.
>
> *Wittgenstein:* Does one want to see that? In the mathematical case, does one want to see what chalk mark the man makes? Surely there is something queer about this.—Does one want to see what he will get if he multiplies, or what he will get if he multiplies correctly—what the right result is?
>
> *Turing:* One can never know that one has not made a mistake.

It was as characteristic of Wittgenstein to seize on the chalk mark as it was of Turing to lay emphasis on the correlation between figures looked up "in such-and-such tables" (perhaps by a Turing machine) and the direction in which a lever swung. The odd thing was, for all the head butting that they did, both were fundamental pragmatists in a way that Alonzo

Church could never have been. Turing might play the role of the defender of logic as a "clean and gentle" discipline, remote from the ugliness of human endeavor and human conflict, yet his own imagination was taking him as far from Hardy's idealism as Wittgenstein's. At the same time, he was not prepared to accept Wittgenstein's dismissal (for example) of the liar's paradox, which lay at the root of Turing's investigations into the *Entscheidungsproblem*, as nothing but a "useless language-game": "If a man says 'I am lying' we say that it follows that he is not lying, from which it follows that he is lying and so on. Well, so what? You can go on like that until you were black in the face. Why not? It doesn't matter." For Turing, it did matter—not in some abstract or ideal sense but because he believed that hidden contradictions could result in things "going wrong." Their argument extended over the course of the entire year and reached its culmination in a long discussion about the role of the law of contradiction in logic and (again) "ordinary" life:

> I may give you the rules for moving chessmen without saying that you have to stop at the edge of the chessboard. If the case arises that a man wishes to make a piece jump off the chessboard, we can then say, "No, that is not allowed." But this does not mean that the rules were either false or incomplete.—Remember what was said about counting. Just as one has freedom to continue counting as one likes, so one can interpret the rule in such a way that one may jump off the board or in such a way that one may not.
>
> But it is vitally important to see that a contradiction is not a germ which shows general illness.
>
> *Turing:* There is a difference between the chess case and

the counting case. For in the chess case, the teacher would not jump off the board but the pupil might, whereas in the counting case we should all agree.

Wittgenstein: Yes, but where will the harm come?

Turing: The real harm will not come in unless there is an application, in which case a bridge may fall down or something of that sort.

The bridge came up (and fell down) again and again. Just as Turing was adamant that Wittgenstein should admit the possibility of a bridge disaster brought on by the misapplication of "a logical system, a system of calculations," Wittgenstein was adamant that Turing should draw a distinction between the world of logic and the world of bridge building. Thus to Turing's repeated assertion that "practical things may go wrong if you have not seen the contradiction," Wittgenstein replied,

> The question is: Why are people afraid of contradictions? It is easy to understand why they should be afraid of contradictions in orders, descriptions, etc., *outside* mathematics. The question is: Why should they be afraid of contradictions inside mathematics? Turing says, "Because something may go wrong with the application." But nothing need go wrong. And if something does go wrong—if the bridge breaks down—then your mistake was of the kind of using a wrong natural law.

Yet Turing would not forget the bridge. It was as if, sitting in Wittgenstein's rooms, he could see it collapsing, hear the cries of the pedestrians as they fell into the river. His point was simple, and he would not let it go: Turing, in his own words, objected "to the bridge falling down."

> *Wittgenstein:* But how do you know that it will fall down? Isn't that a question of physics? It may be that if one throws dice in order to calculate the construction of the bridge it will never fall down.
>
> *Turing:* If one takes Frege's symbolism and gives someone the technique of multiplying in it, then by using a Russell paradox he could get a wrong multiplication.
>
> *Wittgenstein:* This would come to doing something which we would not call multiplying. You give him a rule for multiplying; and when he gets to a certain point he can go in either of two ways, one of which leads him all wrong.
>
> Suppose I convince Rhees of the paradox of the Liar, and he says, "I lie, therefore I do not lie, therefore I lie and I do not lie, therefore we have a contradiction, therefore $2 \times 2 = 369$." Well, we should not call this "multiplication"; that is all.
>
> It is as if you give him rules for multiplying which lead to different results—say, in which $a \times b \neq b \times a$. That is quite possible. You have given him this rule. Well, what of it? Are we to say that you have given him the wrong calculus?
>
> *Turing:* Although you do not know that the bridge will fall if there are no contradictions, yet it is almost certain that if there are contradictions it will go wrong somewhere.
>
> *Wittgenstein:* But nothing has ever gone wrong that way yet. And why has it not?

Wittgenstein appears to have been not unsympathetic to Turing. Indeed, he went to great lengths to make sure that

Turing felt he had his say. Yet his impatience was visceral, and obviously exacerbated by Turing's stubborn refusal to distinguish between the collapse of logic and the collapse of a bridge:

> [*To Turing*] Before we stop, could you say whether you really think that it is the contradiction which gets you into trouble—the contradiction in logic? Or do you see that it is something quite different?—I don't say that a contradiction may not get you into trouble. Of course it may.
>
> *Turing*: I think that with the ordinary kind of rules which one uses in logic, if one can get into contradictions, then one can get into trouble.
>
> *Wittgenstein*: But does this mean that with contradictions one *must* get into trouble?
>
> Or do you mean the contradiction may tempt one into trouble? As a matter of fact it doesn't. No one has ever yet got into trouble from a contradiction in logic. [It is] not like saying "I am sure that that child will be run over; it never looks before it crosses the road."
>
> If a contradiction may lead you into trouble, so may anything. It is no more likely to do so than anything else.
>
> *Turing*: You seem to be saying that if one uses a little common sense, one will not get into trouble.
>
> *Wittgenstein*: No, that is *NOT* what I mean at all.—The trouble described is something you get into if you apply the calculation in a way that leads to something breaking. This you can do with *any* calculation, contradiction or no contradiction.
>
> What is the criterion for a contradiction *leading* you into trouble? Is it specially *liable* to lead you into trouble?
>
> It cannot be a question of common sense; unless *physics*

> is a question of common sense. If you do the right thing by physics, physics will not let you down and the bridge will not collapse.

At one point in the course of the lectures, citing Hilbert, Wittgenstein insisted that his purpose was not to drive his students "out of the paradise which Cantor has created"; rather, it was to compel them to question whether that paradise was really worth staying in in the first place:

> I would say, "I wouldn't dream of trying to drive anyone out of this paradise." I would try to do something quite different: I would try to show you that it is not a paradise—so that you'll leave of your own accord. I would say, "You're welcome to this; just look about you."

Turing, however, had long since abandoned that paradise—which made Wittgenstein's insistence that he play the role of its defender doubly ironic: once again, he was compelled to dress up as someone he was not, and to wear the mantle of a thesis in the validity of which he did not ultimately believe.

3.

Wittgenstein was fond of battle metaphors. "Suppose I am a general and I receive reports from reconnaissance parties," he asked in the twenty-first lecture. "One officer comes and says, 'There are 30,000 enemy,' and then another comes and says, 'There are 40,000 enemy.' Now what happens, or what might happen?"

The imminence of war was clearly on Wittgenstein's mind. "Suppose I am a general and I give orders to two people," he proposed in the next lecture.

> Suppose I tell Rhees to be at Trumpington at 3:00 and at Grantchester at 3:30, and I tell Turing to be at Grantchester at 3:00 and to be at Grantchester at the same time as Rhees. Then these two compare their orders and they find "That's quite impossible: we can't be there at the same time." They might say the general has given contradictory orders.

Turing never received orders to go to Grantchester. Instead, on September 4, 1939, he reported to Bletchley Park, a stately pile in Buckinghamshire, about fifty miles northwest of London. Bletchley Park—or B.P., as it came to be known—had begun its life around the time of the Battle of Hastings, and was a fairly modest red-brick farmhouse until Sir Herbert Leon, a London financier, purchased the estate in 1883. Wanting a mansion grand enough to suit his wealth, Sir Herbert made numerous additions to the building, including an ice house, an entrance hall, a library, and a ballroom. Less judiciously, he added a brick-and-stone façade in what the historian Stephen Budiansky calls "a sort of Victorian mock-Tudor" style, replete with "arches, pillars, gables, domes, and parapets. . . . The interior was equally overdecorated in an unsettling combination of carved oak and red plush." So hideous was Bletchley Park that in an essay entitled "Architecture and the Architect," David Russo gives it as an example of what not to do when designing a house, noting that "even to the untrained eye the structure seems to consist of a variety of forms as though it was not built as a whole but rather built in parts that were later juxtaposed by whim. . . . The resultant building . . . appears to be part castle, part turreted Indian gazebo overlaid with a variety of styles ranging from Romanesque gates to neo-Norman pediments."*

Architectural integrity, however, was not on the mind of Admiral Quex Sinclair (better known as "C") when he bought the house to serve as the base for the continuing activities of the General Code and Cipher School during the Second World War. What attracted him to the property was its spaciousness, its accessibility to London, and its situation exactly midway between Oxford and Cambridge on the railway line then connecting the two universities. Already it was obvious to C that on the cryptanalysis front, at least, the war was going to be fought by intellectuals.

Turing arrived at Bletchley as part of a corps of scientists and mathematicians traveling under the guise of "Captain Ridley's Shooting Party." Among the other members of the group were two fellow Cambridge mathematicians, Gordon Welchman (who later wrote the first memoir of Bletchley Park) and John Jeffreys. Along with a fourth mathematician, Peter Twinn, they were given space in a low building not far from the main house called the Cottage, where they settled down to the task of breaking Germany's Enigma code. Turing probably imagined that his stay at Bletchley would last for a few months; in fact the Crown Inn in the village of Shenley Brook End, where he was billeted and from whence he bicycled each day to Bletchley, was to become his home for the duration of the war.

*In fact, when Michael Apted made the film *Enigma* (a film, incidentally, conspicuous for the absence of any character resembling Turing), he chose to use a different country estate for location shooting.

Back at Princeton, Turing had built an electronic multiplier capable of enciphering messages by multiplying large binary numbers together. Although the idea was crude, it foretold the cryptographic method—based on the multiplication of immense prime numbers—that today protects our credit card numbers when we shop on the Internet. It had no bearing, however, on the German ciphers with which Turing and his colleagues were now presented. Instead, these were highly complex variations on the most basic cipher of all, the so-called *monoalphabetic cipher*.

To understand how a monoalphabetic cipher works, xibauswe rgw xinnib weeie that occurs when one's fingers slip on the typewriter keyboard. "Consider the common error" comes out as "xibauswe rgw xinnib weeie." Now imagine that one is obliged to make sense of a long passage typed with the fingers misaligned: the text appears meaningless. Because we know, however, that the misalignment has resulted in each letter of the alphabet being replaced by another letter, common sense tells us that we should look for repeated three-letter sequences that might represent enciphered versions of the most common word in English, "the." If we were looking at a long stretch of text ciphered according to the finger-slip method, we might notice lots of instances of the sequence "rgw." Assuming, then, that "rgw" is "the," we take the message and replace all the r's with t's, the g's with h's, and the w's with e's. Soon we begin to see other familiar words taking shape. It's rather like an acrostic, which is why monoalphabetic ciphers (which date back to the Middle Ages) are so simple to break. At the heart of the matter is probability—in particular, letter distribution: you use as your guide a knowledge of the statistical frequency with which individual letters appear in

English or whatever other language the plain text is written in. (In English, the most common letter is E, the least common Z.)

Building on the basic principle of substitution, however, one can construct a much more complex cipher, of a sort known as a *polyalphabetic cipher*. A good example is the

		a	**b**	**c**	**d**	**e**	**f**	**g**	**h**	**i**	**j**	**k**	**l**	**m**	**n**	**o**	**p**	**q**	**r**	**s**	**t**	**u**	**v**	**w**	**x**	**y**	**z**
1	**A**	a	b	c	d	e	f	g	h	i	j	k	l	m	n	o	p	q	r	s	t	u	v	w	x	y	z
2	**B**	b	c	d	e	f	g	h	i	j	k	l	m	n	o	p	q	r	s	t	u	v	w	x	y	z	a
3	**C**	c	d	e	f	g	h	i	j	k	l	m	n	o	p	q	r	s	t	u	v	w	x	y	z	a	b
4	**D**	d	e	f	g	h	i	j	k	l	m	n	o	p	q	r	s	t	u	v	w	x	y	z	a	b	c
5	**E**	e	f	g	h	i	j	k	l	m	n	o	p	q	r	s	t	u	v	w	x	y	z	a	b	c	d
6	**F**	f	g	h	i	j	k	l	m	n	o	p	q	r	s	t	u	v	w	x	y	z	a	b	c	d	e
7	**G**	g	h	i	j	k	l	m	n	o	p	q	r	s	t	u	v	w	x	y	z	a	b	c	d	e	f
8	**H**	h	i	j	k	l	m	n	o	p	q	r	s	t	u	v	w	x	y	z	a	b	c	d	e	f	g
9	**I**	i	j	k	l	m	n	o	p	q	r	s	t	u	v	w	x	y	z	a	b	c	d	e	f	g	h
10	**J**	j	k	l	m	n	o	p	q	r	s	t	u	v	w	x	y	z	a	b	c	d	e	f	g	h	i
11	**K**	k	l	m	n	o	p	q	r	s	t	u	v	w	x	y	z	a	b	c	d	e	f	g	h	i	j
12	**L**	l	m	n	o	p	q	r	s	t	u	v	w	x	y	z	a	b	c	d	e	f	g	h	i	j	k

Vigenère cipher, which has its origins in the fifteenth century but came into more frequent use in the late nineteenth century. Essentially, a Vigenère cipher works by means of the construction of a "tableau" in which the plaintext letters are listed across the top and the "key" letters along the left side. (For concision's sake, I have included here a tableau written out only as far as the twelfth position, L; the reader can assume that a full tableau would continue in the same fashion through Z.)

To use this cipher, the sender and recipient need only agree on a keyword. Let's say the keyword is DELILAH and that the message I want to encode is "Be at Grantchester at three." I then replace each letter in my plaintext message with the corresponding letter in the column marked by the appropriate letter in the keyword:

D	E	L	I	L	A	H	D	E	L	I	L
b	e	a	t	g	r	a	n	c	h	e	s
e	i	l	b	r	r	h	q	g	s	l	d

My message now begins EILBRRHQGSLD. . . . With the same keyword, the receiver can decipher it and get the plaintext. This cipher differs from a simple monoalphabetic cipher in that, obviously, each letter is enciphered by use of a different substitution alphabet; therefore the cipher cannot be broken by simply looking for repeating sequences of letters and guessing what they might represent.

But it *can* be broken. The trick, as with all ciphers, is to look for a point of vulnerability and then take ruthless advantage of it until the fortifications collapse. In any polyalphabetic cipher, the obvious weak point is a dependence on repetition; in the example given above, the keyword DELILAH is seven letters long, which means that in a long message, the first, eighth, fifteenth, and twenty-second letters are all being coded with the same monoalphabetic cipher, as are the second, ninth, sixteenth, and twenty-third letters, the third, tenth, seventeenth, and twenty-fourth letters, etc. In principle, then, in order to have a shot at deciphering the message, you would need to have a keyword that was not terribly long; you would need to know its length; and you would need to possess a large enough body of enciphered text to allow for several repetition cycles. You would then break the text into units the length of the keyword and align them in order to study the letter frequencies.

Unfortunately, code breakers rarely have this kind of information to hand. To make matters more complicated for them, as an extra security measure, most users of polyalphabetic ciphers would alter the order of letters within the keyword according to a prearranged scheme; that is, they would agree to begin ciphering one day with HDELILA, the next with AHDELIL, the next with LAHDELI. What was needed was a theoretical leap forward, and not surprisingly, the first mathematician to recognize the most promising route of attack was none other than Charles Babbage, creator of the analytic engine. As Simon Singh explains in *The Code Book*, Babbage's stroke of genius was to step back from the microanalysis of letter frequencies in a sequence of enciphered text and instead treat the sequence as if it had been randomly generated. Essentially Babbage was enacting a statistical analysis of letter repetitions—a method that Solomon Kullback, building on the work of his mentor, William Friedman, would formalize in the article "Statistical Methods," published in 1938 in *Cryptanalysis.*

Here's how it works.* Let's say that we have a sequence that is twenty-four letters long. We can break it into two segments of twelve letters each and then align them. Statistically, the chance, say, that P will appear in position 7 in the first twelve-letter segment is 1 in 26, as is the chance that P will appear in position 7 in the second segment. This means that the chance of P's appearing in position 7 in *both* segments is $\frac{1}{26} \times \frac{1}{26}$, or .15 percent. But the chance of *any* letter's appearing in the same position in both segments is $26 \times \frac{1}{26} \times \frac{1}{26}$, or 3.8 percent. On the other hand, if the two aligned segments have actually been ciphered using the same keyword, the two letters in each position will be part of a monoalphabetic substitution, and the P in position 7 in the first segment will be the enciphered form of the same letter as the P in position 7 in the second segment. In any stretch of English text, E has a 12 percent chance of appearing at any given position, which means that in two aligned segments of text ciphered with the same keyword, the letter that has been substituted for E has a $\frac{12}{100} \times \frac{12}{100}$, or 1.4 percent, chance of appearing in the same position in both segments. A similar percentage can be worked out for each of the other twenty-five letters in the alphabet, and when these values are averaged, you get a rate of coincidence around 6.7 percent. It is therefore possible to shift the alignment of the enciphered text segments in relationship to each other and perform a frequency analysis on each pair. If the rate jumps from 3.8 percent to 6.7 percent, you know that you have stumbled upon two segments

*The principal sources for the account that follows are Singh's *The Code Book* and Budiansky's *Battle of Wits.*

ciphered with the same keyword, and you can proceed from there, applying the same acrostic puzzle method that you would use in the case of a monoalphabetic cipher.

As the weaknesses in the Vigenère cipher and others of its kind became apparent, cryptographers began seeking out refinements by means of which the ciphers they used could be rendered more resistant to cryptanalysis. One approach was to create a keyword the same length as the message; as Singh observes, however, this approach ended up proving only slightly less vulnerable than the original method, since the cryptanalyst could simply fit common words such as "the" into the cipher text at various points, and then see if the substitution generated a segment of key that might likely be a portion of an English word. If testing out the word "the" generated the letters XGT, you could pretty much assume that you were on the wrong track. On the other hand, if the substitution led to DAY, you would know, at the very least, to continue your investigation, since this letter sequence is a common one in English.

A further refinement involved the replacement of keywords by randomly generated key *sequences* of letters that had no meaning in English (or whatever language was the basis for the cipher). This resulted in an unbreakable cipher, but one with the huge disadvantage that its use required the generation of a great quantity of randomly generated letter sequences. In the years before the computer, such sequences were nearly impossible to create. In addition, there was the problem of distributing the key sequences among the operators in the network. In principle this could have been done through the printing of pads, on each page of which a different random sequence would be listed. The operators would use the sequence indicated for a given day, then at the end of the day tear it off and throw it away. Unfortunately, the logistics involved both in printing the pads and in getting them to all the operators in the network, especially during wartime, proved in the end so cumbersome as to render the system more or less unusable.

The next logical step was to build a cipher machine. The first cipher disc, made from copper, was built in the fifteenth century. In the nineteenth century Thomas Jefferson invented a "cipher cylinder," a device consisting of discs printed with letters in different sequences mounted on an axle. A similar "cipher wheel" was used by the Confederacy during the Civil War. These devices, however, merely mechanized the work of feeding a letter through the Vigenère cipher; the cipher text that resulted was no less impervious to attack for having been generated by a machine.

What was needed was a machine that would not only speed up the processes of encipherment and decipherment but actually augment security by subjecting the letters fed into it to extra scrambling. This breakthrough came about in the 1920s, with the more or less simultaneous invention in the Netherlands, Sweden, the United States, England, and Germany of a type of machine of which the German version—the Enigma—would become the most successful exemplar.* The Enigma was the brainchild of the German engineer Arthur Scherbius, who exhibited his device for the first time at the Congress of the International Postal Union in Bern, Switzerland, with the ambition of selling it to businessmen concerned

*Simon Singh notes that none of these machines was very successful, the most spectacular failure being the American version, Edward Hebern's "Sphinx of the Wireless."

that their competitors would intercept their telegrams. As it happened, the Enigma ended up being something of a flop with its target audience: it was both too expensive and, at twelve kilos, too heavy to appeal to bottom-line oriented German entrepreneurs. Several years later, however, the machine came to the attention of the client that would make Scherbius's career: the German government, which bought a slew of the machines and adapted them for military use.

Like that of the personal computer—which in many ways it foreshadowed—the Enigma's apparent simplicity and ease of use belied an ingenious and highly complex internal mechanism. It looked like a typewriter and was no more difficult to use. However, unlike its English cousin, the Typex, the Enigma could not print and had no slot for paper. Instead, mounted above the keyboard, the letters on which were arranged as on a German typewriter, there was an array of twenty-six lamps each labeled with a letter of the alphabet and set up in exactly the same formation as the keyboard itself. Above the array of lamps, in turn, there was a hinged lid fitted with three tiny windows. When you lifted the lid, you would see the three rotors that were the key elements of the Enigma's engineering, each fitted with a movable ring also marked with the twenty-six letters of the alphabet. The three rotors could be removed and rearranged in any of six possible orders.

Operating the machine required no knowledge whatsoever of what went on inside it. Indeed, all that sender and recipient needed to agree upon in advance was the key code, the rotor order, and the ring setting. The idea was that the daily settings would be printed either in books or on sheets of paper collected in pads, copies of which would be distributed to everyone in the sending and receiving network; every morning the senders and recipients would "program" their Enigma with the settings for that day, before proceeding with encryption.

Say you had to send a message. You would first look up the ring setting, rotor order, and three-letter key for the day; next, having fixed the settings, you would move the rotors themselves so that the key—let's say it's ATM—showed through the three windows on the lid. Finally, you would take your message and type it into the Enigma one letter at a time. If the first letter in the message was E, you would type E, the machine would whir and click, and then one of the lamps—let's say the one marked U—would light up. Proceeding in this fashion, you would encipher the entire message, noting down each letter, and then transmit the ciphered message via telegraph or radio in Morse code. The receiver would then set *his* Enigma machine to exactly the same settings, feed in the ciphered message, and—lo and behold—the plaintext would emerge. For this was the genius of the Enigma machine. Its engineering not only guaranteed unparalleled security; it allowed for encipherment and decipherment using the same settings. In other words, if feeding an E into an Enigma programmed to certain settings produced a U, then feeding a U into an Enigma programmed to the same settings produced an E. In terms of its fundamental design, the machine differed little from most of its predecessors, which also relied on a system of rotating disks; what made it unique was that it put the letters fed into it through a battery of extra permutations, and did so with extraordinary speed.

The most essential elements of the engineering were the three rotors, which were arranged in three slots from left to right. On the left and right side of each removable rotor were

The Enigma Machine, employed by the Germans to encrypt classified and sensitive messages during World War II. (HultonArchive/Getty Images)

twenty-six contact points corresponding to the twenty-six letters of the alphabet: the surrounding rings were printed with the letters themselves, in alphabetical order, and could be moved around the rotor, thus altering which letter corresponded to which contact point on a daily basis. (One day, for instance, A might be at contact point 17, the next at contact point 3, and so on.)

Inside each rotor a mass of wires connected the contact points on the right face to those on the left. This wiring, though arbitrary, was fixed; that is to say, even though the rotors could be placed in different orders, all the rotors in all the Enigma machines in the system were wired alike. This insured that rotor 1 on the sender's Enigma would have wiring identical to that of rotor 1 on the recipient's Enigma. And since the rotor order was one of the elements fixed in advance, there was no chance of cross-wiring.

A series of switches connected the rightmost of the rotors to the keyboard. In the commercial Enigma, the letters on the keyboard were linked up with the contact points on the first rotor in the same order as that found on the keyboard; in the military Enigma, however, the wiring had been changed, and one of the first challenges the code breakers faced was to figure out what the new order was. Because the top row of letters on a German typewriter reads QWERTZUIO (as opposed to the QWERTYUIOP found on American typewriters), Dilwyn Knox, one of the first Englishmen to take on the Enigma, referred to the mysterious new letter order as the "qwertzu." Though Knox feared that this order might prove so arbitrary as to defy analysis, to his great surprise, a group of Polish cryptanalysts led by Marian Rejewski quickly determined that in the military model of the Enigma, the Germans had simply connected the letters on the keyboard to the contact points on the first rotor in alphabetical order. This was the first of several instances in which German laziness and lack of imagination ended up aiding the code breakers in their effort to defeat the machine.

When a letter was typed into the Enigma, current flowed from the keyboard into the rightmost rotor, which then shifted one position, thus changing the letter's identity. The current continued through the other two rotors, with a substitution occurring in each position. Next the current entered a "reflector," a half-width disk at the left end of the machine with contacts only on its right side. The reflector connected pairs of letters, replacing the incoming letter with a second one, which would then be sent back through the three rotors for another series of substitutions. Its function was to guarantee that no letter typed into the Enigma could be enciphered as itself; it was also responsible for the Enigma's property of being able to serve as both an encipherment and a decipherment machine.

A last element—incorporated into the military Enigmas—was a "stecker" board, rather resembling an old-fashioned telephone switchboard and located at the base of the machine, with twenty-six jacks (or "steckers") into which cables could be plugged. On these machines the positions of the steckers also had to be agreed upon in advance. Steckered pairs of letters would be swapped both before and after the encipherment process took place: to give an example, P might be steckered as V, which would mean that when the operator of the Enigma typed in a P it would immediately be replaced by a V, which would in turn be put through the encipherment process. Along the same lines, if at the end of the encipherment process a C emerged, it might be replaced by a J, assuming that C and J were likewise steckered. Although in principle all twenty-six letters could be steckered in thirteen pairs, during the early years of the Enigma system only six pairs of letters were steckered; later this number was increased to ten. One of the principal challenges the code breakers faced was determining which the letters were.

From a statistical standpoint, the sheer multitude of scramblings to which the Enigma subjected any incoming plaintext made it close to impregnable. In contrast to the Vigenère cipher, in which the sequence of monoalphabetic ciphers began again with each repetition of the keyword, an Enigma would have to run through the full cycle of all three rotors—a stretch of more than 17,000 letters—before a key sequence would repeat. (Three rotors with twenty-six contact points each meant that there were 26^3, or 17,576, possible routes that a letter could take.) In addition, the rotors could be removed and arranged in any one of the six different orders. Later, the German military added two extra rotors, making it possible to arrange three out of five in any of six orders, for a total of sixty possible rotors orders; later still, the total quantity of rotors was increased to eight. The adjustability of the ring setting allowed for further complication, since the ring on each rotor could be shifted into any of 26 positions, for a total of 676 positions when all three rings were taken into account. Finally, a stecker board consisting of only six cables could generate 100,391,791,500 further permutations.

Not surprisingly, even at the very end of the war, when Turing and his team were cracking messages right and left, the Germans continued to believe that the Enigma system was completely impervious to attack. Instead, they blamed the security lapses they were witnessing on espionage, or the presence of double agents within their own ranks. It was inconceivable to them that an encipherment system so sophisticated as the Enigma's could prove susceptible to interception. After all, it was a *German* machine. Yet, as it turned out, the Enigma was immensely susceptible. Indeed, well before Captain Risley's shooting party arrived at Bletchley Park, the team of

Polish mathematicians led by Rejewski had been reading German military Enigma traffic for several years.

The astonishing saga of the code breakers is really an example of the power of mathematics. Hardy's "clean and gentle" science, as it turned out, was stronger than the entire German war machine, which, for all its posturing, ended up being trumped by a group of geeky mathematicians and engineers working out their ideas on paper and fitting electrical switches inside ugly-looking machines. Luck had something to do with it: the Poles, for example, had at their disposal a commercial Enigma and two purloined German Enigma manuals, each including photographs and instructions. They also had much enciphered Enigma traffic to analyze. Had it not, however, been for the flashes of insight that led Rejewski and his team to crack the code, the intercepted telegraphs might have remained gibberish. Instead, through a combination of mathematical theory and sheer patience, Rejewski—and later Turing—managed to see a way through.

To some degree, their success owed to an innate weakness in the design of the Enigma. On every machine the *rightmost* rotor would have to shift through all twenty-six of its possible settings before the middle rotor made even a single shift; likewise the middle rotor had to shift twenty-six times before the leftmost rotor shifted once. As Budiansky explains, this meant that for stretches of twenty-six letters in any message enciphered using the Enigma machine, the settings to the left of the rightmost wheel remained unaltered: just the sort of weak point on which trained cryptanalysts know to pounce. Had the fast rotor been in the second or third position, the prospects for breaking the code would have been considerably dimmer.

On several fronts, moreover, German efforts to *improve* the security of the Enigma actually ended up making things easier for the cryptanalysts. One example of this kind of lapse was the German decision, early in the war, to instruct everyone in its Enigma network to double-encipher the messages. Upon sending a message, the operator would not use the daily key: instead, he would choose a key at random—say GSX—then encipher that key twice using the daily indicator key—say AMT. The result would be a sequence of six letters (say JMGVEB) that would represent the double encipherment of the key the operator had chosen, GSX, using the indicator AMT. Upon getting the message, the recipient would likewise set his Enigma machine to AMT and feed in JMGVEB, which would come out the other side as GSXGSX, thanks to the Enigma's property of reversibility. The recipient would then reset the Enigma to GSX in order decipher the actual message.

From the standpoint of a German military officer intent on increasing the security of his Enigma traffic, this system of double encipherment must have seemed a stroke of genius; from a mathematical standpoint, on the contrary, it opened up a hole in the system. For the six letters that prefaced each of the enciphered Enigma messages sent on a given day—JMGVEB—in fact amounted to a faint "signature": that is, in every message sent on a given day, the first six letters would have been enciphered using the same three- letter indicator key (in this case AMT).* In effect, this meant that the first six letters of every message being sent were enciphered using the equivalent of a polyalphabetic cipher the keyword of which was only three let-

*As Hodges observes, this practice of incorporating the indicator code into the body of the message paralleled Turing's idea of expressing instructions in the same mathematical language as processes in the universal machine.

ters long, with the first letters all coded using one monoalphabetic cipher, the second letters all coded using a second monoalphabetic cipher, and so on. Moreover, there would be a traceable association between the first and fourth letters of the preface, as there would be between the second and fifth and the third and sixth. It was this point of vulnerability—the repetition in the signature—that Rejewski and his colleagues were able to exploit, since it allowed them to build up chains of letter associations, and by means of those chains to reconstruct all the cipher alphabets necessary for the breaking of each day's traffic. (The creation of letter chains, as it happens, was an innovation of which Turing would take advantage when he assumed charge of the Enigma project a few years later.)

Unfortunately, the Germans soon abandoned this precaution in favor of a simpler method that did not require the use of daily keys. Now the sender of the message would simply choose a key himself (say AGH), then choose a second key (say DJX) with which he would twice encipher the original key. He would then transmit the original key unciphered followed by the twice-enciphered version of the original key. The result would be something like AGHLMODMP. The text of the message would then be enciphered using DJX as the key. The recipient would now set *his* Enigma to AGH in order to retrieve the key necessary to deciphering the message, in this case DJX. The system was not vulnerable to eavesdroppers, because not just the key AGH but also the secret daily ring setting were necessary in order to decipher the key DJX.

The new method of transmitting the key, though simpler, in fact made cryptanalysis more difficult because it erased the "signature" on which Rejewski and his colleagues depended. However, there was a way through. In a statistically significant number of instances, a letter in the key code being transmitted would purely by chance end up being enciphered *twice* as the same letter. For instance, in the example given above, LMODMP represents the double encipherment of DJX: feeding it into an Enigma set at AGH gives us DJXDJX. Note, however, that in this case J has twice been enciphered as M, purely by chance. For reasons lost to history, the Poles referred to these repetitions as "females." Rejewski's colleague Henryk Zygalski now went to work cataloging which of the 105,456 combinations of rotor orders and rotor settings (that is, 17,576 × 6) resulted in females. This task took the better part of a year. Next he crafted a series of perforated sheets in which the punched holes indicated all those positions in which a combination of rotor order and rotor setting resulted in a female. By repeatedly shifting the sheets in relation to one another atop a light table, and noting the points at which the light shone all the way through a perforation, Zygalski was able to work out the ring settings for a given day, and hence make a first step toward deciphering the traffic.

So far the Poles had dealt valiantly with every change that the Germans made to the Enigma system. Their ingenious work culminated in the invention of a machine that Rejewski called the bombe, either because of the ticking sound it made or (a less likely explanation) because he was eating an ice cream bombe at a café when the idea for it hit him. The bombe was capable of simulating the activity of several Enigmas wired together and could run through the 17,576 possible rotor settings of the Enigma in roughly two hours. By November 1938 six Polish bombes were in operation—and then, on December 15, 1938, and January 1, 1939, respectively, the Germans introduced two innovations to the

Enigma that left Rejewski and his fellow mathematicians reeling. First, the Germans added two more rotors to the original complement of three; next they increased the number of steckered letter pairs from six to ten.

The result was more than the Poles could handle: they had neither the manpower nor the money to contend with such staggering new odds, especially given Poland's increasing vulnerability to Germany. At this point, therefore, the principal Allied cryptanalytic effort shifted to Bletchley, where Turing and Welchman had at their disposal not only the resources of the British government but a considerably larger workforce. Before being forced to flee their homeland, however, the Poles were able to deduce the wirings of the two new rotors. They were thus able to present to Turing and Welchman, as a sort of parting gift, the complete wiring of all five rotors.

4.

Although the operations at Bletchley were ostensibly being run under the aegis of the military, the atmosphere that prevailed on the estate was a distinctly casual one. No one wore uniforms. Photographs show the cryptanalysts playing rounders on the lawn in front of the manor house. They took breaks for tea and generally enjoyed the fresh air and bucolic landscape of the local countryside.* And yet there was no

*In the fall of 1941, however, administrative problems, overcrowding, and the poor quality of the Bletchley plumbing, among other concerns, prompted Turing to write, in conjunction with Welchman and two other colleagues, a letter to Winston Churchill asking for help. Churchill's response was swift and emphatic: "Make sure they have all they want on extreme priority and report to me that this had been done."

question but that they understood just how deadly serious their work was. For instance, they knew better than to talk about what they were doing, even with their families, lest somehow the Germans should get wind of the fact that the Allies were reading their Enigma traffic. (For years Alan Turing's mother knew only that her son was involved in some sort of government work.) A certain upper-class English suavity defined the mood of the place, a tacit acknowledgment that no matter how grim the situation got, duty required them to keep working, smiling, and, above all, silent.

And work they did. For security's sake, the labor was divided up among groups each of which was assigned to its own building, most of these wooden "huts" that had been constructed in preparation for the estate's transformation into a code- and cipher-breaking center. In Hut 8, Turing oversaw the theoretical side of things. Other huts were dedicated to the interception of coded traffic, to its transcription, to its translation, and to its interpretation. Crucially, each hut functioned independently of the others, which meant that Turing probably never learned what benefit the messages he deciphered provided the British in their struggle to defeat Hitler. His focus, instead, had to remain on the theoretical and mathematical dilemmas inherent in the effort to break this hugely difficult code.

One refreshing aspect of life at Bletchley was the large number of women employed there, most of them "Wrens" (members of the Women's Royal Naval Service) from Cambridge and Oxford, who operated the decipherment machines and did much of the transcription work. Other women workers were recruited from a nearby corset factory. There was even one woman on the cryptanalytic team, Joan

Clarke, a mathematician to whom Turing became briefly engaged. According to Hodges, when Turing admitted his homosexuality to his fiancée, she was unfazed; shortly thereafter, however, he decided that he could not go through with the marriage and broke the relationship off.

As Singh explains, the advent of the Enigma machine had heralded a substantive change in the science of cryptanalysis. "For centuries," he writes, "it had been assumed that the best cryptanalysts were experts in the structure of language. . . ." Now, however, recruiters were focusing on finding men and women possessed both of striking creative capacities and innate patience. In addition to career mathematicians, the team at Bletchley included the British chess champion, Hugh Alexander, the writer Malcolm Muggeridge, and the winners in a competition to solve the *Daily Telegraph* crossword puzzle as fast as possible. (The record time recorded was 7 minutes, 57.5 seconds.) To succeed at cryptanalysis, one also had to be able to combine mathematical cleverness with a certain instinct for practical application: exactly the recipe that Alan Turing had brought to his efforts to solve the *Entscheidungsproblem* and that had left him feeling, in other contexts, like such an outsider.

At first, Turing and his colleagues at the Cottage emulated the methods of the Poles, creating a series of perforated sheets that could be laid one over the other in various arrangements. When light shone through all the sheets at once, it meant that the cryptanalysts had detected a "female." Then, in order to accelerate the process of hunting for females, they built a small machine that they called, appropriately enough, a "sex cyclometer."

Soon, however, it became clear that the old methods were

not going to be sufficient, especially in light of the changes to the stecker board and the addition of the extra rotors. Instead, an entirely new framework would have to be developed if the team at Bletchley was to succeed in deciphering even a fraction of the Enigma traffic. And this framework Turing (whom his underlings had taken to calling "the Prof") did develop, in an astonishingly short span of time. The result was the "Prof's Book," a messy, on occasion nearly illegible document several hundred pages in length, in which he laid out in detail the theoretical underpinnings of his planned attack on the Enigma. Given the increase in steckering, Turing saw, the cryptanalysts were going to have to depend increasingly on "cribs"—specific segments of plaintext that they could match, with reasonable confidence, to specific segments of ciphertext. As an example of a crib, Turing gave the plaintext (in German) "keine Zusätze zum Vorbericht" ("no additions to the preliminary report"), which corresponded to a stretch of cipher text as follows:

D A E D A Q O Z S I Q M M K B I L G M F W H A I V

K E I N E Z U S A E T Z E Z U M V O R B E R I C H T

The idea was to feed the message containing the crib through the various possible settings at which the ciphering process on an Enigma might begin, and then see which, if any of them, generated a comprehensible plaintext. If none worked, it would be necessary to start again, matching the crib to a different segment of ciphertext. But this was an extremely time-consuming process, as well as one in which, working by hand, cryptanalysts were likely to make mistakes. Moreover, it was not possible to make even a dent in the huge volume of

Enigma traffic by the use of just scissors, pens, and pencils.

Early on in his stay at Bletchley, it had become evident to Turing that the only way to break a cipher created by a machine would be with machine. The insight was a variation on the one that had led him to write "Computable Numbers." This time, however, the machine in question could not remain merely hypothetical. He had to build it.

The result was the second-generation bombe, both faster and more technically complex than its Polish forebear. Also bigger: more than six and a half feet high and seven feet wide, and weighing a ton. In essence, this mechanical behemoth simulated the efforts of thirty Enigma machines working at once. The rotors—ninety of them—were mounted on the face of the immense cabinet, a glance at the back of which revealed more than ten miles of wire connected to contact points on the rotors. The bombe could be temperamental, giving its operators electric shocks or nipping at their fingers. It leaked oil and regularly jammed. But it worked, and eventually a whole series of bombes was commissioned, each one given a unique name. (These included Victory, Otto, Eureka, and Agnus Dei.)

Designing the bombe gave Turing the opportunity, at long last, to fulfill a lifelong dream. As a child he had drawn a blueprint for a typewriter. After writing "Computable Numbers," he had made significant progress toward constructing both the electronic multiplier and the machine to test out the zeros of the Riemann zeta function. But he had never actually completed any of his machines. Now, at Bletchley, he was being given the chance not just to apply principles of mathematical logic to the actual construction of a machine but to oversee its installation and put it to work. For the miracle of the bombe was that this ungainly conglomeration of multiwire cables,

The so-called "bombe," designed by Alan Turing and his colleagues at Bletchley Park to decode encrypted Enigma traffic. (Imperial War College)

brushes, and switches operated entirely according to the methods Turing had learned as a result of his deep immersion in the world of Frege and Russell; indeed, as each bombe clicked its way through thousands of eliminations and checks each day, it was as if the heartbeat of logic itself was being heard.

Yet Turing's achievement went beyond merely building the bombes: along with his colleague Gordon Welchman, he also figured out novel and ingenious ways to use the machine. For instance, one of the principal challenges facing the cryptanalysts at Bletchley at the beginning of the war was the dilemma of how to deal with the millions and millions of new letter combinations that resulted from the increase of steckered letter pairs from six to ten. At first the problem seemed insurmountable; rather quickly, however, Turing came up with a geometric model for chains of letter combinations within the Enigma that swept away the effect of the stecker board entirely. In effect, he took a geometer's approach to the problem.

Here is Stephen Budiansky's example of what Turing did. Let's assume that, by using a crib, we've worked out a clear matchup between a plaintext and a ciphertext. First we lay out the relative positions of the letters:

Relative position	1	2	3	4	5	6	7	8	9
Plaintext	M	I	T	S	C	H	L	A	G
Ciphertext	H	M	I	X	S	T	T	M	I

It is now possible to map the geometric relations between the plaintext letters and the ciphertext letters. For example, in position 6, H is transformed into T, while at position 3 T is transformed to I; at position 2, I is transformed to M, while at position 1, M is transformed to H. We now have a closed loop of letters, from H back to H. Similar closed loops could be mapped for the remaining letters. Using these loops, Turing was able to work up a diagram of the steckering used in any particular message, thus eliminating the stecker board's effect.

Turing's insight, in this case, owed entirely to his mathematical training, from which he had learned that geometric relationships remain constant even as the variables introduced into them are changed. Turing also exploited—cunningly—what were considered the Enigma's greatest strengths: its reversibility, which allowed it to be used as both an encipherment and a decipherment machine, and the fact that it never ciphered any letter into itself. Finally, he built a version of the principle of reductio ad absurdum into the engineering of the machine, which could, in effect, draw conclusions from contradictions:

that is, it was designed to interpret the registering of an invalid rotor setting as an instruction to test out the next possible rotor setting. The machine would stop only when either one circuit or twenty-five circuits were energized, indicating the possibility of a setting that actually worked.

In striving to master the Enigma, Turing and his colleagues took advantage of every bit of outside help they could get. Much of their work rested on an intricate (and brilliant) mathematical foundation; at the same time, they benefited considerably from the lack of imagination that appeared to be endemic to the German military. For instance, getting hold of usable cribs would have proven to be considerably more difficult had the authors of the messages fed into the Enigma bothered either to avoid common phrasings or to ensconce the actual message in nonsense text. More commonly, the messages intercepted by Bletchley were replete with formulaic language, military clichés, and habitual repetitions: they reeked of bureaucracy. (Most messages, for example, mentioned weather conditions, almost always according to the same formula; thus, as Singh notes, *Wetter*—the German word for "weather"—was a common crib.) When the supply of cribs threatened to run dry, the Royal Air Force sometimes helped the cryptanalysts out by planting mines in locations specifically chosen so that the Germans would find them and send back reports of the discoveries; because the English already knew where the mines were, these location reports provided ready-made cribs that Turing and his team could exploit. This process was called, rather quaintly, gardening.

Another German error of which Turing took full advantage was the fact that in choosing three-letter message keys, German operators often took the path of least resistance. An operator might, for instance, repeatedly choose the first three letters of his girlfriend's name; or he might resort to the use of three-letter sequences that cut diagonals across the typewriter keyboard: RFV and TGB, say. The cryptanalysts knew to look for such repetitions, which they called cillis, or sometimes sillies, and when they found them (which was often), their work was made that much simpler.

Lastly, the mathematicians at Bletchley got considerable help from the British navy, which periodically managed to confiscate Enigma equipment, codebooks, logbooks, and instruction manuals from sunken U-boats and intercepted German trawlers. On one occasion, the gunner on an English ship managed to salvage a waterlogged canvas bag containing two days worth of Enigma settings as well as an operator's log including full plaintext and ciphertext for all messages sent on those days. On another, sailors boarding a U-boat located two intact Enigma rotors: this was a particularly helpful catch, because it came just after the Germans had increased the number of rotors from five to eight for naval Enigma traffic. The two rotors were, in fact, two of the three that had just been added.

But for every piece of good luck, there was a setback. Occasionally the Germans would make unexpected changes in their method of transmitting messages (usually for the purpose of increasing *internal* security and protecting against spies), thus wiping out weeks of intense cryptanalytic labor and forcing the team at Bletchley to go back to the drawing board. While certain Enigma networks operated according to principles that rendered their traffic easier to read, moreover, one system—the naval Enigma—employed a host of extra security measures that made it maddeningly resistant to interception. For instance, the naval Enigma not only had a provision for the use of three out of eight rotors (increasing the number of possible rotor orders by a factor of almost six); it also employed an adjustable reflector that could be fixed in any of twenty-six positions. (Later the German navy started using a *four*-rotor Enigma machine.) To make matters even more complicated, operators in the naval Enigma system dispensed altogether with the old system of beginning each message with a ciphered version of the key necessary for the message's decipherment; instead, they coded the key using an entirely different system based on bigram substitution tables; two three-letter groups selected by the operator would each be padded with a fourth random letter, then aligned:

P J L O

M Q B A

The letters in the group would then be arranged as vertical "bigrams," or groupings of two letters:

PM JQ LB OA

The bigrams, in turn, would be replaced with those indicated on a daily bigram substitution table. The apparent impenetrability of the naval Enigma was particularly problematic because its traffic contained the information that the British admiralty needed most urgently in order to secure the waters of the Atlantic and the Pacific against U-boat attacks.

On occasion a seemingly inconsequential change in the method of transmitting Enigma traffic—such as an alteration to the bigram substitution tables—was enough to stop the cryptanalysts at Bletchley in their tracks for weeks at a time. Fortunately they soldiered on; indeed, their eventual victory over the Enigma owed as much to the hours of drudge work put in by the many very dedicated (and very patient) men and women who worked at Bletchley as it did to the theoretical and engineering breakthroughs for which Turing himself was chiefly responsible. Through tireless analysis of the mathematical foundation of polyalphabetic ciphers, relentless exploitation of even the tiniest chinks that appeared in the machine's armor, and the clever use of electrical equipment to mechanize and thus dramatically speed up the testing and elimination of thousands upon thousands of possible key combinations, Turing managed to render the Enigma, if not a powerless antagonist, at least a manageable one. Eventually he even cracked the naval Enigma code, thus bringing about a sharp decrease in the sinking of Allied ships by U-boats. It was a bravura, even heroic performance, which contributed significantly to Hitler's defeat.

It was also a performance of which Turing was the undisputed organizer. True, the original conception of the bombe had to be credited to the Poles. Yet, as Budiansky points out, "the fundamental mathematical insight behind the British bombe was wholly Turing's," as was "the discovery that matching strings of plain and cipher text defined a characteristc geometric relationship" and "the idea of feeding a contradiction back into an interconnected loop of Enigma machines." As Hodges notes, Turing got the last laugh on Wittgenstein, since "*these* contradictions would make something go very wrong for Germany, and lead to bridges falling down."

5.

It was during his years at Bletchley that Turing earned his reputation for eccentricity, social awkwardness, and slovenliness.

The episode of his aborted enrollment in the home guard dates from this period. So do certain legends that, true or not, still circulate: that he wore a gas mask when he bicycled to work each morning (supposedly to keep from breathing in pollen); that when he rode his bicycle he counted the revolutions of the wheels, stopping one revolution before he knew the chain was going to come off; that at the end of each day he chained his tea mug to a radiator pipe. At Bletchley, rumor had Turing keeping his trousers held up with string, wearing his pajamas under his sports coat, rarely shaving or cutting his nails. All of which might have been true: he had never been very tidy. Such atypical conduct as he displayed, moreover, could be easily written off as actually characteristic of a certain "type": the "absent-minded professor" of which Sidney Stratton was another exemplar. And yet simply to laugh off Turing's idiosyncratic behavior during his years at Bletchley is to miss both its more disturbing and less comic implications. Yes, tea mugs were in short supply during the war, but didn't chaining one to a radiator suggest a slight degree of paranoia? Likewise the obsessive counting of tire revolutions on a bicycle he never had repaired, since if he had, as Hodges observes, it would have meant that someone else would have been able to ride it. Turing might have no use for the conventions (bucking clichés of masculinity, he learned to knit while at Bletchley), nor any patience for "pompousness or officialdom. . . . [H]e wouldn't suffer fools or humbugs as gladly as one sometimes has to." Still, he himself had acknowledged the necessity of subscribing, at least to some degree, to the social niceties, when he had written to his mother at Princeton that his friend Maurice Pryce was "much more conscious of what are the right things to do to help his career." Sworn to secrecy regarding his work, bereft of pretty much all sexual possibility, and sidetracked by necessity from the work on the universal machine for which he had been preparing before the war, Turing seems to have gone beyond the pale at Bletchley, gradually losing whatever capacity he possessed for playing by the rules.

As the war drew to a close, the work Turing had at first orchestrated was able to move forward more or less by itself. In deciphering Enigma traffic, Bletchley was meeting with a degree of success of which even the most overconfident of cryptanalysts would scarcely have dreamed. The running of the bombes became an industrial matter; "Job up! Strip machines," the operators would cry once the Enigma cipher for the day had been broken. Yet as the well-oiled machine of code breaking began, as it were, to run itself, its architect was left with increasingly little to do. Bletchley sent him to the United States, to consult on the construction of an electronic bombe intended to deal with the threat of a four-rotor naval Enigma. He also contributed to the construction of a machine intended to decipher an entirely different type of traffic, enciphered on a teleprinter and known as Fish. In a more immediate sense, however, his work as a code breaker was finished.

The extent of his contribution to the war effort—of which he never spoke during his lifetime—should not be underestimated, and though it would probably be an exaggeration to say that without Turing the Allies would not have won the war, it is reasonable to suppose that without him it would have taken them several more years to do it. At the same time, had the British authorities known that Turing was homosexual, they might have refused to let him anywhere near

Bletchley, in which case, as his friend Jack Good observed, "we might have lost the war."

Turing's years at Bletchley constitute the best-documented period in his life, yet in the end his work as a code breaker amounted to a very long diversion from his dream of building a universal machine. For the bombes were about as far from universal as you could get. Their very design guaranteed their obsolescence, since it depended on the quirks and particularities of another, much smaller machine, the Enigma, of which the bombe was the huge, distorted shadow. Nor did the high stakes of the venture, the pressure to decipher as much Enigma traffic as swiftly as possible, allow Turing any latitude or free time to experiment with the possibility of setting a universal machine to the specific task of code breaking. Speed was too much of the essence. Because, in "Computable Numbers," the machine of which Turing wrote was intended for use only in the landscape of the theoretical, its fastness or slowness was inconsequential. But during the war there was no time to waste. Lives depended—literally—on how well the bombe was able to do its job.

Today Bletchley Park continues on, as a sort of museum and memorial to the men and women who gave so many years to breaking the Enigma code. There's a terrific gift shop at which you can buy Enigma key chains, T-shirts, and wall magnets. Former Wrens, now in their eighties, take you on a tour through the mansion, the outbuildings, and the various huts, including Hut 8, which now contains a huge replica of one of the bombes. Presumably, the originals were destroyed after the war ended, both for the sake of security and because they no longer served any purpose. They were as onetime as the onetime pads used in sending a cipher.

Turing's presence is everywhere. The guides, as they lead you through, talk warmly about "the Prof." They show you with pride the ornamental oars from a 1935 "bumps" race at King's College, displayed in the main house and engraved with the news that Turing replaced W. M. "Bill" Colles on the number two boat.

During a lull in my tour of Bletchley, as we were approaching a monument to the Polish code breakers, I told my guide that I was writing a book about Turing. She shook her head and said, "What a tragedy. People didn't really understand about homosexuality in those days."

The implication, of course, was that today people understood better.

"Did you know him?" I asked.

"Oh yes," she said. "Sometimes he tied string around his waist to keep his trousers up."

But then, before I could ask more, the crowd of visitors stilled. My guide cleared her throat and began to tell us the amazing story of Marian Rejewski.

6

The Electronic Athlete

1.

In the summer of 1942, Max Newman, Alan Turing's mentor at Cambridge, arrived at Bletchley, where he was assigned to the analysis of the Fish traffic. Turing's contribution to the project had been the development of a statistical process known as Turingismus. (The mathematicians at Bletchley were fond of monikers of this sort; a procedure developed by Turing as part of the attack on the naval Enigma was nicknamed Banburismus, since the sheets on which he and his colleagues entered their data were printed in the town of Banbury.) Efforts to break the Fish traffic by hand took place in a section of Bletchley called the Testery, but with Newman's arrival, the focus shifted. Like the Enigma code, Fish was generated by a German machine, the Lorenz. Now Newman concluded, as Turing had before him, that the only way to break this machine-generated cipher was with a machine built specifically for that purpose. So he settled down to work in Hut 11, which came to be known as the Newmanry. In his research, he got assistance from the engineers at the recently established Telecommunications Research Establishment, or TRE, in Malvern, as well as the Post Office Research Station at Dollis Hill, in the London suburbs.

The first of the machines designed to take on the Lorenz was delivered by the Post Office to Bletchley in June 1943. It was called the Heath Robinson, in honor of the Edwardian cartoonist whose drawings featured immensely complex industrial contraptions that performed absurdly simple or simply absurd tasks. (Subsequent Robinsons included a Peter Robinson and a Robinson and Cleaver, named after a London department store.) Unfortunately, the name Heath Robinson turned out to be prescient: the machine was notoriously bad-tempered, prone to breaking down and catching on fire. Worse, the teleprinter tapes that were central to its design tended to tear. Luckily, Newman soon found an ally in the electronic engineer Tommy Flowers, who was based at Dollis Hill, and with Flowers's help he was able to overcome the technical difficulties that plagued the Robinsons. The result was the much more efficient Colossus, which employed 1,500 electronic valves and with which Newman successfully tackled the Fish traffic.

Given that its architect was his former mentor, one would think that Turing would have played a central role in the development of the Colossus. As the war neared its end, however, Turing had begun to move away from cryptanalysis and into other areas of research. In addition to his stay in Washington, his trip to the United States had included a visit to Bell Labs in New Jersey, where he had spent two months studying the relatively new science of speech encipherment: essentially, the manipulation and deliberate distortion of sound waves by a signal that would function as a key or indicator and that could be applied both by the sender and by the receiver of the message, scrambling or unscrambling as the case might be. The theory behind speech encipherment was not dissimilar to the one according to which the Enigma had been designed, the difference being that, instead of writing, it was speech itself that was being rendered unintelligible. It was an idea that fascinated Turing; especially after so many years spent searching out and exploiting the vulnerabilities in someone else's cipher, he relished the opportunity to apply all that he had learned to the design of a truly impregnable system of his own.

His exposure to the activities at Bell Labs was not the only source of the liberation that Turing experienced during his months in the States. He also met with Claude Shannon, an American pioneer of computer science based at MIT, with whom he discussed at length the question of whether a machine could be built that imitated the behavior of the brain; Shannon, who believed that one could, went so far as to imagine a day when humans might read poetry to machines or play music for them. Turing also spent a considerable amount of time in New York City, where he professed astonishment at the casual ease with which a stranger propositioned him at a hotel—a different sort of liberation.

Having acquired a good working knowledge of both the electronic equipment and the mathematical theory required to design a speech encipherment system, he returned to Bletchley in the spring of 1943, where he found that Hut 8 had hardly missed him. The chess champion Hugh Alexander, a more able and ambitious manager, had more or less taken over work on the naval Enigma. Watching the bombes tick away as busy Wrens reset and reshuffled the rotors, Turing must have felt like a parent whose child had passed into adolescence and no longer needed him. In addition, he was not particularly keen to join the team at the Newmanry, given that most of the work on the Colossus had been done in his absence. Although Turingismus provided the theoretical and statistical basis for the design of the Colossus, the machine was not his "baby" in the way that the bombe had been. Besides, his imagination was taking him in a new direction.

It was also taking him away from Bletchley. Not terribly far away—only about ten miles north—was another country estate, Hanslope Park. The house dated from the late eighteenth century and since 1941 had served as the base for the secret service's "Special Communications Unit No. 3." Bletchley was becoming overcrowded, and given its proximity, it seemed natural that Turing should set up shop at Hanslope, where, along with his young assistants Robin Gandy (who later wrote extensively about Turing) and Donald Bayley he went to work on the speech encipherment project, offering, in typical Bletchley fashion, a prize to whoever among his colleagues could come up with the best name for it. Gandy won with Delilah, a reference to the biblical temptress who deceived Samson.

The atmosphere at Hanslope was much more formal than that at Bletchley. For one thing, there was a much more visible military presence. Then too, the powers that be did not accord Turing any special privileges. Instead, he was given space in a large hut in which a wide variety of research programs was being carried out. Once again, his own contribution to Delilah put less emphasis on the hardware than on the establishment of a sound theoretical foundation for the system; it was as if, at every opportunity, he was determined to prove

Wittgenstein wrong about the falling bridge, by making sure each step was shored up by logic. Most of the nuts-and-bolts work—literally—he left to Bayley, a youth from the Midlands who had recently graduated from Birmingham University with a degree in electrical engineering; what interested Turing was the theory behind the machine, which he was determined to make as impressive for its simplicity as for its invulnerability. Not for the first time he saw the advantages of applying the aesthetic standards that Hardy had set forth for mathematical proof to the much less rarefied business of building things.

At last Turing moved out of the Crown Inn, taking lodgings first at the officers' mess at Hanslope and then at a cottage in the kitchen garden of the estate, which he shared with Gandy and a cat called Timothy. He took to running long distances, read a lot of Trollope and Jane Austen, and went to parties at the mess—the first time he had had anything like a normal social life since his stay in Princeton. With the end of the war in sight, it was no longer necessary to practice the sort of social austerity to which Turing had become acclimated at Bletchley. Once again, people were allowed to have personal needs. He had never been reticent about his homosexuality. Indeed, even at a time when thousands of Englishmen led outwardly heterosexual lives while secretly engaging in "acts of gross indecency," Turing had displayed a remarkable degree of self-confidence and comfort in his own sexual identity. That he saw his sexuality as part of his identity in the first place put him at odds with the prevalent thinking of his age and reflected, no doubt, the years that he had spent in the privileged corridors of King's College.

Not that he was in any way a zealot. Indeed, in contrast to that of Edward Carpenter, the mathematician turned philosopher whose ideal of male comradeship had inspired Forster to write *Maurice*, Turing's openness owed less to any conscious decision process than to an allergy to dishonesty that was an outgrowth of his notorious literal-mindedness. Put simply, Turing could keep a secret only when he thought there was a good reason to do so. In the case of Enigma, there was obviously a host of good reasons not to tell anyone about the job he'd performed. So far as his homosexuality was concerned, he saw no point in dissembling. (Lyn Newman recalled that Turing "found the idea of deceiving others so distasteful that he supposed it equally so to almost everyone.")

So he told people. He told Joan Clarke. He even told Don Bayley, his assistant at Hanslope. As Hodges describes it, there was nothing somber or earnest about the conversation. He did not sit Bayley down and announce that he had something grave or even consequential to impart. Rather, he just let the news slip casually, while they were working. Bayley's reaction—a frank revulsion perfectly in keeping with his Midlands upbringing—took Turing completely aback.

According to Hodges' account of the incident, what appalled Don Bayley was not merely the fact of Turing's homosexuality, which could be seen as part and parcel of his general eccentricity; it was that Turing "seemed to think it perfectly natural and almost to be proud of it." That he refused to demonize himself, however, did not mean that other people wouldn't demonize him: this was what he appears to have failed, or perhaps even refused, to understand. Forster, less credulous and, generally speaking, more pessimistic, feared that *Maurice* might provoke a similar backlash, and therefore chose not to publish the novel during his lifetime. "If it had ended unhappily, with a lad dangling from a noose or with a suicide pact, all would be well," he wrote in a 1960 terminal note to the novel. Carpenter, he added later in the same note, "had hoped for the generous recognition of an emotion and for the reintegration of something primitive into the common stock. And I, though less optimistic, had supposed that knowledge would bring understanding. We had not realized that what the public really loathes in homosexuality is not the thing itself but having to think about it." It was presumably "having to think about it" that was so upsetting for Bayley.

The problem, in part, was loneliness. Despite his ease with his homosexuality, which did indeed verge on pride, Turing had never had a really fulfilling relationship with another man. Instead, his erotic life so far had consisted of bouts of unrequited longing, usually for heterosexual men who had no interest in him, alternating with occasional "friendships with benefits" with other gay men in whom he had a minimal sexual interest, and with whom he was far from in love. These friendships, by their very nature, were compromises. Better than nothing they might be, but they paled in comparison with the unfulfilled ideal that was Christopher Morcom. It seems not improbable that when Turing let slip the fact of his homosexuality "accidentally," especially to a young man like Bayley, he was hoping against hope that the admission might provoke an expression of reciprocal desire. That rarely happened. Later he told Robin Gandy, "Sometimes you're sitting talking to someone and you know that in three quarters of an hour you will either be having a marvellous night or you will be kicked out of the room." With Bayley, he was very firmly kicked out of the room; indeed, he considered himself lucky that Bayley agreed to continue working with him at all.

By the spring of 1945 the Delilah was operational. It came too late to be of any practical use in the war—which was perhaps part of the reason why the Post Office showed so little enthusiasm for it. (Another reason was that the output was crackly.) Soon enough the bugs had been dealt with, but by then, as was typical for Turing, he had lost interest in the project. For another idea had seized him—or perhaps it would be more accurate to say that it had reseized him. Delilah was a single-purpose machine; the bombe was something even less general, a machine built for the specific purpose of defeating another machine. Both of them were rather like the friendships that in Turing's life had served as stand-ins for the love affairs he had never known. Now he wanted the real thing: to build a machine that was not just universal but to which, as Claude Shannon had speculated, one might read a poem or play a piece of music: a machine that could actually be said to think.

2.

In June 1945 Turing accepted a post as temporary senior scientific officer at the National Physics Laboratory in Teddington, a suburb of North London abutting Bushey Park, at a salary of £800 per annum. Since 1938 the laboratory had been under the directorship of Sir Charles Galton Darwin (1887–1962), the grandson of the father of evolutionary theory and himself an applied mathematician from Cambridge whose field of expertise was x-ray crystallography. Darwin's initiatives at the laboratory included the institution of a new mathematics division, the superintendent of which, J. R. Womersley, had been given the mandate to start a program of research into "the possible adaptation of automatic telephone equipment to scientific computing" as well as the development of an "electronic counting device suitable for rapid computing."

Part of what motivated Darwin and Womersley was a fear that the United States had pulled ahead of Britain in computer research. That very year, at the University of Pennsylvania's Moore School of Engineering, a computer called the ENIAC (short for electronic numerical integrator and calculator) was being put into operation. The brainchild of John Mauchley and J. Prosper Eckert, whose later lives would be marred by legal battles over its patent, the ENIAC employed 17,468 vacuum tubes (as opposed to the Colossus' 1,500) as well as 70,000 resistors, 10,000 capacitors, and 5 million soldered joints. Speed was its principal objective, as the patent application made clear:

> With the advent of everyday use of elaborate calculations, speed has become paramount to such a high degree that there is no machine on the market today capable of satisfying the full demand of modern computational methods. The most advanced machines have greatly reduced the time required for arriving at solutions to problems which might have required months or days by older procedures. This advance, however, is not adequate for many problems encountered in modern scientific work and the present invention is intended to reduce to seconds such lengthy computations. . . .

In other words, the ENIAC was intended to be mainly a very fast number cruncher—not the sort of machine, one assumes, that would respond with much enthusiasm to a sonnet or a sonata. Moreover, it departed radically from Turing's ideal of a universal machine in that it was pretty much all hardware, which meant that in order to change the programming one had literally to open the machine and reattach its thousands of switches and cable connections. To borrow a phrase of Turing's, making alterations to the ENIAC necessitated "screwdriver interference" rather than "paper interference." Turing, on the other hand, envisioned a machine whose hardware would be as streamlined as possible and which one could adapt to different purposes simply by changing its instruction tables.

It was at this point that John von Neumann reentered the picture. Like Turing, von Neumann (or Johnny, as his friends called him) had spent the war years as a consultant to the military, particularly on the development of the hydrogen bomb. He had also served as a member of the Scientific Advisory Committee to the Ballistic Research Laboratories at the Aberdeen Proving Ground in Maryland. All this was work that required large-scale computation of just the sort for which the ENIAC (on which von Neumann consulted) was designed. At the same time, von Neumann recognized the limitations of the ENIAC; with his background in logic, he envisioned a machine less dependent on engineering, more flexible in terms of programming, and—perhaps most crucially—possessing an enormous memory. The machine was to be called the EDVAC (electronic discrete variable automatic computer), and on June 30, 1945, a proposal for its design was delivered to the U.S. Army Ordinance Department under von Neumann's name.

It is not too much of a stretch to say that Turing's fingerprints are all over the report. For instance, of the memory, von Neumann writes,

> While it appeared that various parts of the memory have to perform functions that differ somewhat in their nature and considerably in their purpose, it is nevertheless tempting to treat the entire memory as one organ, and to have its parts even as interchangeable as possible for the various functions enumerated above.

As Hodges observes, von Neumann's "one organ" is pretty much equivalent to Turing's "tape"; indeed, this idea that he finds "tempting" lies at the very heart of "Computable Numbers." Likewise, the EDVAC had a stored program, as opposed to the ENIAC's cable-based programming system. Yet the EDVAC report contains not a single mention of Turing's name. If, as von Neumann claimed, he had indeed never read another paper in logic after his unhappy encounter with Gödel in Königsberg, then his reiteration of many of Turing's key points not only in the report on the EDVAC but in several articles from the period was yet another remarkable example of two mathematicians making the same discovery years apart.

Between von Neumann and Alonzo Church, it appeared that Turing had made so little of an impression at Princeton that he might as well never have been there. Yet if Turing resented von Neumann's apparent wholesale appropriation of his ideas, he said nothing about it, at least publicly. Instead, he focused on the differences and, having settled in at Teddington, went to work on a proposal of his own. This was for a computer that would be called the ACE, short for "automatic computing engine." (The use of the word "engine" might have been meant as an allusion to Babbage's analytical engine.) Turing's report on the ACE, published in 1945 and meant to be read, according to Turing, in conjunction with von Neumann's on the EDVAC, is much more densely detailed than von Neumann's, with logical circuit diagrams and a cost estimate (£11,200). It also posited a machine that was in many ways more radical, and certainly more minimalist, than the EDVAC—not to mention most of the computers in operation today.

What made the ACE unique, in Turing's words, was its capacity to "tackle whole problems. Instead of repeatedly using human labour for taking material out of the machine and putting it back in at the appropriate moment all this will be looked after by the machine itself." Self-sufficiency, however, was only one of several facets of the machine's character that distinguished it from predecessors such as the ENIAC.* It was also to be much less dependent on hardware:

> There will positively be no internal alterations to be made even if we wish suddenly to switch from calculating the energy levels of the neon atom to the enumeration of groups of order 720. It may appear somewhat puzzling that this can be done. How can one expect a machine to do all this multitudinous variety of things? The answer is that we should consider the machine as doing something quite simple, namely carrying out orders given to it in a standard form which it is able to understand.

*The ACE also used stacks to implement subroutine calls, as do all modern machines. The EDVAC did not. Another innovation not in the EDVAC report is Turing's "Abbreviated Computer Instructions," an early form of programming language. I am indebted to Prabhakar Ragde for pointing out these distinctions.

The machine, in other words, would be capable not only of "looking after" itself but of "understanding" instructions. Already Turing's language bestowed personhood upon it. This personhood was not meant to be taken merely as a metaphor or even, as Keynes might have put it, a "state of mind": rather, it owed entirely to the ACE's independence from "screwdriver interference," from the fact that instructions fed in from outside were what gave it identity. Indeed, in a lecture that he delivered on the ACE to the London Mathematical Society on February 20, 1947, Turing went so far as to suggest that, much as a child matures in response to social stimuli and education, such a machine might be capable of growth: "Possibly it might still be getting results of the type desired when the machine was first set up, but in a much more efficient manner. . . . It would be like a pupil who had learnt much from his master, but had added much more by his own work. When this happens I feel that one is obliged to regard the machine as showing intelligence."

The ACE could in theory "learn by experience"—but only if certain technical requirements were met. First, its memory, if not "infinite," would need "to be very large." Another "desirable feature" would be that "it should be possible to record into the memory from within the computing machine, and this should be possible whether or not the storage already contains something, i.e. the storage should be *erasible* [*sic*]." But what form should such storage take? Alluding to "Computable Numbers," Turing rejected his old idea of an "infinite tape" on the grounds that too much time would have to be spent "in shifting up and down the tape to reach the point at which a particular piece of information required at the moment is stored. Thus a problem might easily need a storage of three million entries, and if each entry was equally likely to be the next required the average journey up the tape would be through a million entries, and this would be intolerable." What was needed was a "form of memory with which any required entry can be reached at short notice." To have a "really fast machine," Turing concluded, "we must have our information, or at any rate a part of it, in a more accessible form than can be obtained with books. It seems that this can only be done at the expense of compactness and economy, e.g. by cutting the pages out of the books, and putting each one into a separate reading mechanism. Some of the methods of storage which are being developed at the present time are not unlike this." Wittgenstein, of course, had begun his first lecture on the philosophy of mathematics by imagining a scenario in which Turing, asked to point out a Greek sigma in a book, "cuts out the sign [Wittgenstein] showed him and puts it in the book." To be sure, the philosopher added, such "misunderstandings only immensely rarely arise—although my words might have been taken either way." Now Turing was imagining pages cut out of books and then inserted, in a similar fashion, into "separate reading mechanisms." It was as if he were determined to challenge Wittgenstein, once again, by envisioning a situation in which logic demanded the very kind of literal-mindedness Wittgenstein had mocked.

For Turing was, as ever, literal-minded—so much so that he built a certain literal-mindedness into his design for the ACE: "The machine interprets whatever it is told in a quite definite manner without any sense of humour or sense of proportion. Unless in communicating with it one says exactly what one means, trouble is bound to result." He might have been writing about some of his own more tortured efforts to determine whether another man would be amenable to a caress or a kiss, whether a conversation was going to lead to a "marvellous night" or to being "kicked out of the room." Trouble resulted when channels got crossed—an idea Turing would explore in a later essay. For now his main goal was to ask that his machines be given a fair chance, that they not be faulted simply *because* they were machines. To illustrate this point, he referred to his solution to the *Entscheidungsproblem*, noting that a machine developed to distinguish provable formulae from unprovable ones would sometimes, by necessity, fail to provide an answer. By contrast, a mathematician, upon being given such a problem to solve, "would search around and find new methods of proof, so that he ought eventually to be able to reach a decision about any given formula." Against such an argument, Turing wrote,

> I would say that fair play must be given to the machine. Instead of it sometimes giving no answer we could arrange that it gives occasional wrong answers. But the human mathematician would likewise make blunders when trying out new techniques. It is easy for us to regard these blunders as not counting and give him another chance, but the machine would probably be allowed no mercy. In other words then, if a machine is expected to be infallible, it cannot also be intelligent.

If Turing did not yet understand what it meant to be shown "no mercy," he would, alas, learn that lesson all too soon. For the present, he was content to make a "plea" that his machines be treated with greater tolerance than he, as a homosexual man, was destined to experience:

> To continue my plea for "fair play for the machines" when testing their I.Q. A human mathematician has always undergone an extensive training. This training may be regarded as not unlike putting instruction tables into a machine. One must therefore not expect a machine to do a very great deal of building up of instruction tables on its own. No man adds very much to the body of knowledge, why should we expect more of a machine? Putting the same point differently, the machine must be allowed to have contact with human beings in order that it may adapt itself to their standards.

It all came down to loneliness. Maurice Pryce, who knew better than Turing "the right things to do," was getting on in his career. Other friends were marrying, having children, while Turing had become very much the "confirmed solitary" Newman had feared he might turn into. Now it seemed as if he was determined that the ACE—which had not yet even been born—should have a very different adolescence from his own; that it should enjoy "human contact," and not be subjected to prejudicial injustices. There is in the lecture a clear sense of identification with the machine, as well as a certain protective affection for it—as if, having never found a life's companion, the engineer in him was now determined to build one.

3.

Turing's speculative musings on the ACE—in particular on its prospects for the future—sometimes reveal the same edge of paranoia that marked much of his odder behavior at Bletchley. For instance, late in the lecture before the London

Mathematical Society, he divided those whom he saw as being destined to "work in connection with the ACE" into masters and servants, the masters being the theorists who decide its uses, the servants the technicians who "feed it with cards as it calls for them," keep it in good working order, and assemble its data. As time goes on, however, "the calculator itself," he hoped, could "take over the functions both of masters and servants," with the latter in particular being "replaced by mechanical and electrical limbs and sense organs." The risk, Turing imagined, was that the machine, by virtue of its very capacities, should come to pose a threat to human beings, who would in turn enter into a conspiracy against it (and perhaps, by association, its inventor)—a scenario not unlike the one that befalls Sydney Stratton in *The Man in the White Suit*. But the machine, so far, existed only in the form of a report: in prophesying such a dark and "dangerous" future for it, Turing was not just jumping past its construction, its testing, and its installation; he was taking its total success for granted. He was also putting the possibility of its dangerousness into the heads of the very listeners whose anxieties he was supposed to be trying to assuage.

The problem was not that he lacked the technical know-how necessary to design a computer; on the contrary, the report puts forward an astonishingly detailed battery of specs for the ACE's design. The problem was that, lacking Maurice Pryce's sense of social savoir faire, Turing was inclined to undermine his own chances of winning support by letting his imaginative flights of fancy get the better of him. As always, he said exactly what he believed—and suffered the consequences. For instance, in his lecture before the London Mathematical Society, he argued that it was much more crucial that the ACE be digital than that it be electronic. "That it is electronic is certainly important," he told his listeners, "because these machines owe their high speed to this, and without the speed it is doubtful if financial support for their construction would be forthcoming. But this is virtually all there is to be said on that subject." Speed, of course, had been the ENIAC's raison d'être, while the ACE, with a pulse rate of a million a second, promised to be the fastest machine ever built. Yet, if Turing was to be believed, he was making the ACE fast only to appease the money bags on whose generosity its development depended. That he should have adopted such a patronizing tone not just toward the very people he needed to make the ACE a reality but to the ideal of speed itself was all the more bizarre, given the years he had spent trying to outrace the Germans at Bletchley. No experience could have made Turing more aware of the real value of a fast machine than the long days he had spent laboring to crack naval Enigma messages in time to prevent U-boat attacks.

Both the ENIAC and the EDVAC were also digital machines. The ENIAC, however, performed its calculations using decimal as opposed to binary notation, which bogged down the process, while the EDVAC put its emphasis on number crunching.* Since Turing, by contrast, imagined the ACE being put to many uses that did not involve number crunching, he designed it to be rather minimalist, with the emphasis on programming: in modern parlance, on software rather than hardware. Indeed, in his report Turing listed, among the many possible tasks for which the ACE might be employed, one of interest to the military

*This was also true of the so-called Johnniacs—the von Neumann–style machines that the EDVAC later inspired.

("Construction of range tables"); one of interest to pure mathematicians ("Given two matrices whose coefficients are polynomials of degree less than 10, the machine could multiply the matrices together, giving a result which is another matrix also having polynomial coefficients"); one of interest to engineers ("Given a complicated electrical circuit and the characteristics of its components, the response to given input signals could be calculated"); and one of interest to municipal governments ("To count the number of butchers due to be demobilised in June 1946 from cards prepared from the army records"). He also listed a function of interest to children (doing a jigsaw puzzle) and one of interest to himself (playing chess). The ACE's digital nature meant that it could "be made to do any job that could be done by a human computer, and . . . in one ten-thousandth of the time." And this was to a great extent *because* the machine was so remarkably simple, employing only the most basic symbolic vocabulary: "To perform the various logical operations digit by digit, it will be sufficient to do 'and,' 'or,' 'not,' 'if and only if,' 'never' (in symbols *A & B*, $A \vee B$, $\sim A$, $A \equiv B$, *F*)." More complicated arithmetic—even addition and subtraction—would be part of the programming, which again put the machine at odds with the EDVAC, in which the arithmetic was to be performed by feeding numbers into the machine's accumulator.

Its digitalness, however, was not enough to guarantee the ACE's success. The implementation of an efficient storage system was also central. In the lecture, Turing listed the three types of storage that in his view would best serve a machine such as the ACE: magnetic wire, cathode-ray tubes, and acoustic delay lines. Cathode-ray tubes, or even iconoscopes of the sort used in televisions, were, he felt, "much the most hopeful scheme, for economy combined with speed." But they were not yet widely available in England, so he opted for the delay lines, which guaranteed the ACE the capacious memory that it would need if it was going to be independent. The ultimate point of having a big memory was to allow the computer's operators—the servants—to forget the more tedious aspects of programming, which the machine would take care of by itself.

Clarity and concision were of paramount importance for the servants, who would presumably work the machine in ignorance of its engineering: "It should be possible to describe the instructions to the operator in ordinary language," Turing wrote, "within the space of an ordinary novel." And the language, for the sake of the servants, *was* ordinary, if a little morbid. One put away subsidiary tables, for instance, by "burying" them, a task one achieved through the use of "a standard instruction table BURY"; likewise, one fetched the tables by "disinterring" them through the use of "the table UNBURY." Not that the servants could *create* these ur-tables; instead, they would "have to be made up by mathematicians with computing experience and perhaps a certain puzzle-solving ability. . . . This process of constructing instruction tables should be very fascinating. There need be no real danger of it ever becoming a drudge, for any processes that are quite mechanical may be turned over to the machine itself."

It was a scenario not unlike the one at Bletchley, where masters who had won crossword puzzle contests worked out the theory in one hut, while in another, Wren-like servants performed the day-to-day chores involved in running, and looking after, the beloved child: the machine.* What remained unspecified was the role that Turing—as inventor, father, and lover—was expected to play.

4.

Turing finished the report on the ACE in 1945 and gave it to Womersley, who in turn brought it before the executive committee of the National Physics Laboratory (or NPL) on March 19, 1946. The ACE's creator also spoke at the meeting. His presentation did not go over terribly well—Darwin in particular appeared not to "grasp the principle of universality," while Turing lost many members of the committee by letting his talk become overly technical. Nonetheless, in the end Darwin recommended that Turing be allocated £10,000 for the construction of a smaller version of the ACE—a "Pilot ACE." Had he won the committee's full confidence, he probably would have gotten more money, but a Pilot ACE was better than no ACE, and he accordingly set up shop in Teddington and went to work.

It was a transitional moment not just for him but for England. Yes, the war was over, but what was to happen next? And what would the future hold for Alan Turing, whose huge

*In her memoir Mrs. Turing recalled, "The most that Alan told me about his war work was that he had about 100 girls under him. We knew one of these 'slaves' as he called them. From her came the information that they marvelled at her temerity in greeting him on Christmas morning with 'A Happy Christmas, Alan,' for they held him in great awe, largely because when he rushed into their part of the building on business, he never gave the least indication that he even noticed them. The truth probably is that he was equally alarmed by them."

contribution to the effort at Bletchley remained (and would presumably remain for years) shrouded in official secrecy? Few at Teddington had a clue as to how much they owed him—a state of affairs that only intensified his sense of leading a solitary, ciphered existence. At Bletchley he had taken to long-distance running. Now he joined the Walton Athletic Club, the membership of which, Mrs. Turing noted in her memoir, "comprised men from all walks of life—road-sweepers, clergymen's sons, dentists, clerks and so forth—he was always at ease among them and made them feel at ease." By way of practice, he frequently ran the eighteen miles from Hampton-on-Thames, where he lived in a guesthouse called the Ivy House, to his mother's house in Guildford. Likewise, when he needed to go to the Post Office laboratories at Dollis Hill, he ran the fourteen miles, usually wearing old flannel trousers tied at the waist with rope.

Perhaps thanks to its independence from the war effort, the atmosphere at Teddington was more bureaucratic and less encouraging of collaboration than at Bletchley, with a clear division between the engineers and the theorists. Turing was expected to function as an idea generator and leave the building to the engineers. At first the NPL made a point of publicizing its support for him. In an interview with the BBC, Darwin cast Turing as a sort of boy wonder (this even though he was already in his midthirties), explaining that "about twelve years ago a young Cambridge mathematician, by name Turing, wrote a paper which appeared in one of the mathematical journals, in which he worked out by strict logical principles how far a machine could be imagined which would imitate processes of thought." From this arcane publication, Darwin implied, the possibility had dawned of a technological miracle from which ordinary Englishmen would benefit. Not surprisingly, such a mythology appealed immensely to the popular press, in particular the tabloids, which soon began calling Turing for interviews. His mother recalled one "evening paper" going so far "as to head a short paragraph about Alan with the words 'Electronic Athlete.'" The emphasis was usually on the astonishing "feats" that the new "electronic brain" would be able to perform, in particular feats of memory to match those of any music hall memory whiz. For instance, the *Surrey Comet* quoted Turing as saying that the ACE would "be able quite easily to remember about ten pages from a novel, though not, of course, in their ordinary form. They would have to be translated into a medium it is capable of 'understanding,' in other words into the digits that it is designed to handle." Similarly the ACE would at least in theory be able to play an average game of chess, though whether it could ever develop the "power of judgment" needed to play a good game of chess remained "a matter for the philosopher rather than the scientist."

Though the "electronic brain" might be the darling of the *Daily Telegraph*, at the NPL it was beginning to cause some worry. Since news of the ENIAC had broken, there had been extensive information sharing between the Americans and the British. Womersley traveled to the States, as did Turing himself. Moreover, von Neumann's influence had started to make itself felt at Teddington: if the EDVAC represented the direction that computer research was bound to take, would the British be foolish to follow Turing's hunch and plan for such a different kind of machine? Would they end up being left behind? Or should the NPL hedge its bets? Notably, Maurice V. Wilkes, a former classmate of Turing's and now the director of the Mathematical Laboratory at Cambridge, had attended a course of lectures in Philadelphia in the summer of 1946 sponsored by the group at the Moore School that had put together the ENIAC. Excited by what he had learned, Wilkes had returned eager to begin work on constructing a British version of the EDVAC. Although Wilkes intended the project to be based in Cambridge, he sought the cooperation of the NPL in putting together a plan for a machine that bore a much closer resemblance to the EDVAC than to the ACE. As if to underline the similarity, Wilkes's computer was to be called the EDSAC—the electronic delay storage automatic computer—and it quickly won both the approbation and the attention of the NPL.* Turing's ACE might fascinate the popular press, but it was out of line with the mainstream. In addition, its projected cost was skyrocketing.

In part, the problem, once again, was Turing's deafness to the conventions. Although the press might cast him as a sort of Chatterton of the computer world—a "marvellous boy"—his peers knew better: proper Englishmen did not tie their trousers with rope. Nor did they jog to meetings in Dollis Hill. His insistence on going his own way was typical; now, however, he was asking England to take it on faith that it should follow him. And England balked.

At first it appeared that the NPL was going to give equal support to both projects—Wilkes, after all, had funding from Cambridge already—and when the report on the proposed EDSAC was completed, Womersley made a point of asking Turing to read it and give his opinion. Turing was not much

*Wilkes was later credited with the invention of microprogramming, for which he received the Turing Award, the Association for Computing Machinery's highest honor for lifetime achievement.

impressed. "I have read Wilkes' proposal for a pilot machine," he wrote to Womersley on December 10, "and agree with him as regards the desirability of some such machine somewhere. . . . The 'code' which he suggests is however very contrary to the line of development here, and much more in the American tradition of solving one's difficulties by means of much equipment rather than by thought." No doubt Turing hoped that by appealing to the NPL's nationalistic pride, he might ensure its continued support. Unfortunately, his tendency to run off at the mouth in interviews was proving to be a source of some embarrassment to the laboratory's board. Womersley suggested that, rather than speaking with reporters, Turing should give a course of lectures on the ACE "intended primarily for those who will be concerned with the technical development of the machine." Not surprisingly, Wilkes, who attended the lectures, complained that Turing was "very opinionated" and that his ideas "were widely at variance with . . . the main stream of computer development."

They were—and remain so. Today the ancestry of most of the computers that we use can be traced back to the EDVAC—and not to the ACE, which in the end never even got built. Though Turing's "minimalist ideas" were in Martin Davis's words "destined to have little or no influence on computer development," their legacy can still be felt in microprogramming, "which makes the most basic computer operations directly available to the programmer"; in the advent of the silicon microprocessor, which is in effect a universal machine on a chip; and in the "so-called RISC (reduced instruction set computing) architecture," which "uses a minimal instruction set on the chip, with needed functionality supplied by programming." All of these owe much to the ACE.

The saddest part of the story, at least in Davis's view, is the degree to which, for years, Turing got written out of the history of the discipline that he effectively invented. Though the Pilot ACE, for instance, survived, Turing had long since left Teddington by the time it actually got built, at which point it had gone through so many redesigns that it bore little resemblance to the machine of which he had dreamed. More cruelly, a 1949 report asserted that "the actual size of the ACE as originally contemplated was the outcome of long consideration by Mr Womersley and Professor von Neumann during Mr Womersley's visit to the United States." According to the evolutionary principles espoused by Sir Charles Darwin's grandfather, it was perhaps inevitable that the soft-spoken Turing should end up being bulldozed by the debonair Johnny von Neumann. Indeed, as late as 1987, Davis reports, when he published an article claiming that von Neumann had gotten many of his ideas from Turing, he felt himself "to be very much alone" in that point of view. Davis was therefore gratified when twelve years later *Time* magazine not only named Turing one of the twenty greatest scientists of the twentieth century but in its entry on von Neumann (who also made the list) wrote,

> Virtually all computers today from $10 million supercomputers to the tiny chips that power cell phones and Furbies, have one thing in common: they are all "von Neumann machines," variations on the basic computing architecture that John von Neumann, building on the work of Alan Turing, laid out in the 1940s.

5.

For Turing the 1940s was an era defined more by beginnings than by culminations. Ideas would come to him, he would throw himself into them, and then, before he had brought them to fruition, he would drift away, either because circumstances compelled him to or because some other idea had seized his attention. Thus by the time Don Bayley presented an operational version of the Delilah before the Cipher Policy Board in 1945, Turing had already left the project, moving on to Teddington, and the ACE. Likewise, he had left Teddington by the time the Pilot ACE was tested. At Bletchley, when his colleagues had talked about what they planned to do after the war, he had always said that he intended to resume his fellowship at King's. On September 30, 1947, he did just that. Officially he was taking a sabbatical—the idea was that at Cambridge he would do theoretical work that he could later apply to the building of the ACE—but in fact both he and Darwin probably knew that he would never return to the NPL.

According to Mrs. Turing, her son decamped to Cambridge because he was "disappointed with what appeared to him the slow progress made with the construction of the ACE, and convinced that he was wasting time since he was not permitted to go on the engineering side." It was a relief to find himself back in bookish and tolerant Cambridge, where he could once again work as he liked. Another plus was that Robin Gandy was now at Cambridge, where he had become a member of the Apostles. Once again, Turing was not elected to this society, which had mattered so much to Forster, and of which Forster had written in *The Longest Journey*. However, he did join the play-reading Ten Club, the Moral Science Club, and the Hare and Hounds Club, under the aegis of which he was able to continue running. He also began a relationship with Neville Johnson, a third-year mathematics student, that would last for several years—again, not a love affair so much as a "friendship with benefits."

To some degree, at Cambridge Turing really was able to take up as if the war years had never happened, and in 1948 he published two papers in mathematics journals: "Rounding-off Errors in Matrix Processes" in the *Quarterly Journal of Mechanical and Applied Mathematics* and "Practical Forms of Type Theory" in Church's *Journal of Symbolic Logic.* He also played chess with the economist Arthur Pigou, who recalled that his opponent "was not a particularly good player over the board, but he had good visualizing powers, and on walks together he and an Oxford friend used to play games by simply naming the moves. This, from the point of view of a chess master, is very small beer . . . but for us humble wood-pushers it was impressive." According to Pigou, Turing "was interested in many other things" besides mathematics "and would gallantly attend lectures on psychology and physiology at an age when most of us were no longer capable of sitting on a hard bench listening to someone else talking."

Another important friendship formed during the sabbatical was with Peter Matthews, then in his second year of the natural sciences tripos, with whom Turing discussed the relationship between physiology and mathematics. Turing introduced Matthews "to the similarities between computing engines and brains," a comparison that Matthews found "very useful." Appositely, on January 22, 1948, Turing gave a talk to the Moral Science Club on "Problems of Robots."

Most of his year at Cambridge, however, Turing devoted to trying to decide what to do with himself once the year was over. One option was to remain at King's, resume the career of a pure mathematician that the war had interrupted, and hope for a lectureship. Another was to return to the NPL—as he was officially supposed to—and continue working on the ACE. A third (and this was, to him, perhaps the most attractive of the alternatives) was to take a position at Manchester University, where since 1946 Max Newman had been in residence as Fielden Professor of Pure Mathematics. Building on the work he had done on the Colossus, Newman was collaborating with the electrical engineer F. C. Williams to develop a computer to rival the EDSAC. Working with Tom Kilburn, Williams had developed a storage system based on the cathode-ray tube that was proving to be much more efficient, flexible, and reliable than the EDSAC's mercury delay lines. The Williams-Kilburn tube, as it came to be called, displayed information as dot patterns and also allowed for the first true use of random-access memory in the history of computer design. Williams later recalled (a bit inaccurately, since Turing had at this point not yet joined the project):

> With this store available, the next step was to build a computer around it. Tom Kilburn and I knew nothing about computers, but a lot about circuits. Professor Newman and Mr. A. M. Turing . . . knew a lot about computers and substantially nothing about electronics. They took us by the hand and explained how numbers could live in houses with addresses and how if they did they could be kept track of during a calculation.

Newman's plan, as he laid it out in a letter to von Neumann, was for a machine that could take on "mathematical problems of an entirely different kind from those so far tackled by machines . . . , e.g. testing out (say) the 4-colour theorem* or various theorems on lattices, groups, etc. . . ." On a philosophical level the kind of inquiry he had in mind was much more in line with Turing's interests than the speed-for-speed's-sake ethos that governed the ENIAC. Nor was money a problem: Newman had a Royal Society grant worth £20,000 to cover the cost of construction, plus £3,000 a year for five years.

It was becoming increasingly obvious at Manchester that the NPL strategy of maintaining a rigid division between the engineering and the mathematical arms of computer development was destined to prove totally counterproductive; intellectual synergy depended not just on allowing ideas to be shared but on recognizing that the barrier erected by the NPL was completely arbitrary. Nor was this view held only at Manchester. At Cambridge, Wilkes was moving ahead with the EDSAC in an environment likewise marked by collaboration between engineering and mathematics. He also had his own money.

It seems likely that Wilkes and Turing distrusted each other. Although Wilkes's laboratory was only minutes from King's, for months Turing avoided going to visit him there. When he finally did go, all he could say was that Wilkes looked like a beetle. Yet if Turing felt envious of Wilkes, he had every right. Not only did Wilkes have a more secure position—as well as the support of the university—he had the ear

*The "four-color theorem," proven in 1997, holds that if you color in a map divided into discrete regions (such as a map of the counties in the state of Florida), you will need a minimum of four colors if no two adjoining regions are to be the same color.

of the NPL, where Womersley was becoming increasingly disenchanted with Turing's maverick and minimalist design. Fearing rightly that the NPL might once again end up being left behind, Womersley now made enquiries as to the progress of the Manchester computer while simultaneously proposing to Darwin that the NPL team use "as much of Wilkes' development work as is consistent with our own programming system" in overhauling the ACE. Soon enough nearly everything that made the machine unique, and uniquely Turing's, would be erased from its design, as the ACE was normalized, brought into line with industry standards.

Not surprisingly, Turing decided to go to Manchester. Newman wanted him there and promised him the chance to do the sort of original research that the culture of the NPL subtly discouraged. More importantly, he would be in on the ground floor on the development of a machine that was actually going to be built—and built in an atmosphere decidedly more sympathetic than the one in Teddington. In May 1948, therefore, Turing resigned from the NPL, irritating Darwin, who felt that Newman had stolen his boy wonder away from him. (It appears not to have occurred to Darwin that he might not have done much to make the boy wonder want to stay.) Before he took up his new post, however, Turing wrote a last report for the NPL. It was entitled "Intelligent Machinery," and it would prove to be one of the most startling, even subversive, documents in the history of computer science.

6.

Like many of Turing's later papers, "Intelligent Machinery" mixes hard-core technical analysis with passages of philosophical, sometimes whimsical speculation. At the heart of the paper is a discussion of the possibility that "machinery might be made to show intelligent behavior." Before delving into this discussion, however, Turing gives a list of what he sees as the five most likely objections to it: "an unwillingness to admit the possibility that mankind can have any rivals in intellectual power"; "a religious belief that any attempt to construct such machines is a sort of Promethean irreverence"; "the very limited character of the machinery which has been used until recent times (e.g. up to 1940)," which has "encouraged the belief that machinery was necessarily limited to extremely straightforward, possibly even to repetitive, jobs"; the discovery, by Gödel and Turing, that "any given machine will in some cases be unable to give an answer at all," while "the human intelligence seems to be able to find methods of ever-increasing power for dealing with such problems 'transcending' the methods available to machines"; and, lastly, the idea that "in so far as a machine can show intelligence this is to be regarded as nothing but a reflection of the intelligence of its creator."

Turing's strategy of opening with a summary of the claims of the naysayers foreshadows the gay rights manifestos of the 1950s and 1960s, which often used a rebuttal of traditional arguments against homosexuality as a frame for its defense. He acknowledges from the outset the futility of trying to talk a zealot out of his zealotry, noting that the first two objections, "being purely emotional, do not really need to be refuted. If one feels it necessary to refute them there is little to be said that could hope to prevail, though the actual production of the machines would probably have some effect." The third objection he dispatches by pointing out that existing machines such as the ENIAC or ACE "can go on through immense numbers (e.g. $10^{60,000}$ about for ACE) of operations without repetition, assuming no breakdown," while he dispenses with the fourth by reiterating a point made in his lecture before the London Mathematical Society, that infallibility is not necessarily "a requirement for intelligence." This idea he underscores by means of an anecdote from the life of Gauss:

> It is related that the infant Gauss was asked at school to do the addition 15 + 18 + 21 + . . . + 54 (or something of the kind) and that he immediately wrote down 483, presumably having calculated it as (15 + 54) (54–12)/2.3.* One can imagine circumstances where a foolish master told the child that he ought instead to have added 18 to 15 obtaining 33, then added 21, etc. From some points of view this would be a "mistake," in spite of the obvious intelligence involved. One can also imagine a situation where the children were given a number of additions to do, of which the first 5 were all arithmetic progressions, but the 6th was say 23 + 34 + 45 + . . . + 100 + 112 + 122 + . . . + 199. Gauss might have given the answer to this as if it were an arithmetic progression, not having noticed that the 9th term was 112 instead of 111. This would be a definite mistake, which the less intelligent children would not have been likely to make.

*Turing has slightly altered the story to suit his argument. According to the original account, Gauss's teacher asked him to add up all the numbers between 1 and 100. His strategy for coming up with the correct answer—5,050—was to divide the hundred numbers in question into the pairs 1 + 100, 2 + 99, 3 + 98, 4 + 97, etc., thus creating fifty pairs, each of which would add up to 101. Gauss then multiplied 50 × 101, obtaining the correct answer.

Educability, then, is the principal ingredient of intelligence—which means that in order to be called intelligent, machines must show that they are capable of *learning*. The fourth objection—"that intelligence in machinery is merely a reflection of that of its creator"—can thus be countered by recognizing its equivalence to "the view that the credit for the discoveries of a pupil should be given to his teacher. In such a case the teacher would be pleased with the success of his methods of education, but would not claim the results themselves unless he had actually communicated them to his pupil." The student, on the other hand, can be said to be showing intelligence only once he has leaped beyond mere imitation of the teacher and done something that is at once surprising and original, as the infant Gauss did. But what kind of machine would be able to learn in this sense?

By way of answering that question, Turing first divides machines into categories. A "discrete" machine, by his definition, is one whose states can be described as a discrete set; such a machine works by moving from one state to another. In "continuous" machinery, on the other hand, the states "form a continuous manifold, and the behaviour of the machine is described by a curve on this manifold." A "controlling" machine "only deals with information," while an "active" machine is "intended to produce some very definite continuous effect." A bulldozer is a "continuous active" machine, just as a telephone is "continuous controlling" one. The ENIAC and the ACE, by contrast, are "discrete controlling," while a brain is "continuous controlling, but . . . very similar to much discrete machinery." Though "discrete controlling" machines, moreover, are the most likely to show intelligence, "brains very nearly fall into this class, and there seems every reason to believe that they could have been made to fall genuinely into it without any change in their essential properties." Such a classification of the brain as a neural machine neatly reverses the popular conception of the computer as an electronic brain, just as Turing's subtle use of the passive "could have been made" furthers the report's quiet anti-Christian agenda by recasting God as an inventor or programmer whose failure to make *brains* "discrete controlling" was more or less accidental. Had God been a little smarter, Turing implies, he would have designed the brain better.*

Indeed, at this point in the report, one begins to get the sense that Turing's ambition is as much to knock mankind off its pedestal as to argue for the intelligence of machines. What seems to irk him, here and elsewhere, is the automatic tendency of the intellectual to grant to the human mind, merely by virtue of its humanness, a kind of supremacy. Even the science of robotics, on which he spoke at Cambridge before the Moral Science Club, comes in for some mockery, thanks to its emphasis on modeling machines on human beings:

> A great positive reason for believing in the possibility of making thinking machinery is the fact that it is possible to make machinery to imitate any small part of a man. That the microphone does this for the ear, and the television camera for the eye are commonplaces. One can also produce remote-controlled robots whose limbs balance the body with the aid of servo-mechanisms. . . . We could produce fairly accurate electrical models to copy the behaviour

*That Turing considered his ideas anti-Christian is borne out by the title he gave to a talk he delivered in Manchester in 1951, "Intelligent Machinery, a Heretical Theory."

> of nerves, but there seems very little point in doing so. It would be rather like putting a lot of work into cars which walked on legs instead of continuing to use wheels.

And yet if one were to "take a man as a whole and try to replace all the parts of him by machinery," what would the result look like? A latter-day Frankenstein monster, to judge from the description and scenario that follows:

> He would include television cameras, microphones, loudspeakers, wheels and "handling servo-mechanisms" as well as some sort of "electronic brain." . . . The object, if produced by present techniques, would be of immense size, even if the "brain" part were stationary and controlled the body from a distance. In order that the machine should have a chance of finding things out for itself it should be allowed to roam the countryside, and the danger to the ordinary citizen would be serious. Moreover even when the facilities mentioned above were provided, the creature would still have no contact with food, sex, sport and many other things of interest to the human being. Thus although this method is probably the "sure" way of producing a thinking machine it seems to be altogether too slow and impracticable.

Better, perhaps, to design the sort of machine that would please another machine: a brain without a body, possessed at most of organs allowing it to see, speak, and hear. But what could such a machine do? Turing lists five possible applications. It could play games (chess, bridge, poker, etc.), it could learn languages, it could translate languages, it could encipher and decipher, and it could do mathematics.

In fact, over the years computers have been shown to be notoriously resistant to learning languages. On the other hand, they can be very good at games, cryptography, and mathematics—the poetry, as it were, of their language. If they are to undertake these efforts of their own volition, however—if they are to play (and win) at tic-tac-toe, generate an unbreakable cipher, or calculate the zeros of the zeta function—they have to be taught. And who is to teach them? What will be the methods by which the "masters" program into them the ability to learn? Turing's answer to this question (which is really the central question of "Intelligent Machinery") says as much about his own education as about his tendency to think of the ACE as a child—and a British child, at that:

> The training of the human child depends largely on a system of rewards and punishments, and this suggests that it ought to be possible to carry through the organizing with only two interfering inputs, one for "pleasure" or "reward" (R) and the other for "pain" or "punishment" (P). One can devise a large number of such "pleasure-pain" systems. . . . Pleasure interference has a tendency to fix the character, i.e., towards preventing it changing, whereas pain stimuli tend to disrupt the character, causing features which had become fixed to change, or to become again subject to random variation.

This rather draconian theory of child rearing suggests the degree to which Turing had internalized the very ethos of "spare the rod, spoil the child" so dominant in England at the time, and to a version of which he would in just a few years fall victim. Perhaps he'd picked up its rudiments in the psychology classes he'd sat in on during his sabbatical year at King's. Or perhaps he was simply recapitulating the educational principles of Sherborne and other English public schools.

> If the untrained infant's mind is to become an intelligent one, it must acquire both discipline and initiative. So far we have been considering only discipline. . . . But discipline is certainly not enough in itself to produce intelligence. That which is required in addition we call initiative. This statement will have to serve as a definition. Our task is to discover the nature of this residue as it occurs in man, and to try and copy it in machines.

Discipline and initiative: Turing sounds, here, like a headmaster delivering a speech at the beginning of the term. In the background one hears the Duchess's lullaby from Lewis Carroll's *Alice in Wonderland.* The difference is that the Turing Primer on How to Raise a Computer requires not just beating the computer when it sneezes but providing the pepper to make him sneeze. In the experimental "pleasure-pain systems" that he outlines, for instance, the pepper comes in the form of a random sequence of numbers to which the computer is required to respond in certain ways by a table of behavior. What Turing calls "pleasure" and "pain" are in fact merely instructions to perform arithmetical operations, with the pain stimulus canceling all tentative entries and the pleasure stimulus making them permanent. The problem is that in the absence of *any* stimuli, "the machine very soon got into a repetitive cycle. This became externally visible through the repetitive B A B A B. . . . By means of a pain stimulus this cycle was broken." Emotion is thus privileged in the programming process itself, raising the question of whether Turing has imposed arbitrary emotional signs onto a mechanical (and therefore feelingless) process, or whether perhaps human emotion is at heart far more mechanical than we are inclined to admit. Such a strategy fits well with Turing's larger effort, in the report, to demystify the human body by describing it in the most mechanistic language possible, as in this comparison of the nerve and the electrical circuit: "Certainly the nerve has many advantages. It is extremely compact, does not wear out (probably for hundreds of years if kept in a suitable medium!) and has a very low energy consumption. Against these advantages the electronic circuits have only one counter-attraction, that of speed."

The extent to which we ascertain that another entity's behavior shows intelligence, Turing writes at the end of the report,

> is determined as much by our own state of mind and training as by the properties of the object under consideration. If we are able to explain and predict its behaviour or if there seems to be little underlying plan, we have little temptation to imagine intelligence. With the same object therefore it is possible that one man would consider it as intelligent and another would not; the second man would have found out the rules of behaviour.

By way of illustrating this point, he proposes an experiment that foreshadows what would later come to be known as the Turing test. Two rather poor chess players—A and C—are placed in separate rooms between which some system of communicating moves has been established. Meanwhile a third man—B—is operating a machine that has been programmed to play chess. C plays a game with either A or the machine operated by B. Will he be able to guess which one is his opponent? Turing suspects that he will find it quite difficult to tell the difference and concludes by remarking parenthetically that the experiment is one he has actually performed. He does not give the result, however, and so "Intelligent Machinery" ends with a number of questions hanging: can a machine, educated through a system of reward and punishment, be said to be able to think? Are children, when they cry or laugh, revealing some spark of soul that distinguishes them from machines, or simply following "rules of behavior" with which we as spectators empathize because we are familiar with them? Or to put it another way, does asking whether computers think require us to ask, as well, whether humans compute?

7

The Imitation Game

1.

The Manchester in which Alan Turing settled in the fall of 1948 was as noteworthy for its industrial ugliness as for its bad weather. Manchester University, just outside the city center, was equally depressing. In Newman's laboratory, the walls were covered with brown tiles in what F. C. Williams, his partner in the project, called a "late lavatorial" style. Most of the faculty lived in the suburb of Hale, where Turing rented rooms before buying his first and only house, in 1950, on Adlington Road in Wilmslow, Cheshire. These rooms likely resembled one that W. G. Sebald described in *The Emigrants*, "carpeted in a large floral pattern, wallpapered with violets, and furnished with a wardrobe, a washstand, and an iron bedstead with a candlewick bedspread."

The machine on which Turing went to work was a preliminary model intended for small-scale experiments, and thus christened (in keeping with Turing's educational program) the Baby. It had the distinction, however, of employing Williams's and Kilburn's cathode-ray tube technology, which meant that for the first time both the instructions fed into the machine and the results it spit out could be *seen*. Not that the Baby employed anything so sophisticated as a screen: instead, the numbers appeared in form of bright spots on the monitor tubes themselves. Spots, or "bits," were arranged on each tube in a 32 × 32 grid (for a total of 1,024 bits), with each bit charged to represent a 0 or 1. A metal pickup plate was set up to detect the charge, and thus "read" the bit's value. Each 32-bit line in the grid, in turn, represented either a number or an instruction; later, the lines would be lengthened to 40 bits each, with each addressable line containing either one 40-bit number or two 20-bit instructions. As Turing remarked in the programmer's handbook that he prepared for the Manchester computer, the information in the electronic store could be compared "to a number of sheets of paper exposed to the light on a table, so that any particular word or symbol becomes visible as soon as the eye focusses"—an analogy that recalls the perforated sheets employed at Bletchley in the effort to break the Enigma code.

One of the oddities of working with the Manchester computer was the programming notation with which, as Martin Campbell-Kelly puts it, Turing "saddled users of the machine. . . . Each program instruction consisted of 20 bits, which Turing wrote down as four characters using the 5-bit Post Office teleprinter code. In effect he used the teleprinter code as a base-32 number system. . . ." This in turn required Turing to invent a 32-symbol "alphabet" of number equivalencies in which most numbers were paired with letters—9 was D, for instance; 19 was W—while some were represented by symbols (@ for 2, " for 27, £ for 31) and 0 was represented by a slash (/). "Because zero was represented by the forward-stroke character," Campbell-Kelly explains, "and this was the most commonly-used character in the written form of programs and data, one early user decided this must be an unconscious reflection of the famously dismal Manchester weather as the effect was that of rain seen through a dirty window pane!" (*//////////////*) As if things weren't complicated enough, numbers entered into the machine had to be written backward. Using the base-32 code, the 40-digit binary sequence 10001 11011 10100 01001 10001 11001 01010 10110 (in denary notation, 17 27 5 18 17 19 10 13) would thus have to be written as Z"SLZWRF—which would, of course, first have to be reversed. This had the effect of leaving anyone who wished to use the machine—including Turing's assistants, Audrey Bates and Cicely Popplewell—rather beholden to its language teacher. Indeed, when Turing delivered a lecture on "Checking a Large Routine" at Cambridge on June 24, 1949 (the day after his thirty-seventh birthday), his failure to bother to clarify the notational system in which he was writing figures on the blackboard struck Maurice Wilkes, who was in the audience, as "bizarre in the extreme. . . . [Turing] had a very nimble brain himself and so no need to make concessions to those less well-endowed." The base-32 code was rather like the bicycle that Turing had had at Bletchley, rigged up so that no one but he could ride it.

By way of an experiment to test the efficiency of the Baby, Newman decided to put to it one of the great puzzles of pure mathematics. This involved the so-called Mersenne primes, named after the French monk Marin Mersenne (1588–1648), who in 1644 undertook an investigation into the interesting fact that certain large prime numbers take the form $2^n - 1$ where n is also prime. As Mersenne soon discovered, the rule did not hold for all prime *n*'s. (For instance, $2^{11} - 1$ isn't prime, though 11 is.) However, by the nineteenth century it had been shown that the rule did hold when *n* was equal to 2, 3, 5, 7, 13, 19, 31, 67, and 127. In 1876 Edouard Lucas (1842–1891) came up with a method by which $2^{127} - 1$ was shown to be prime, and in 1932 D. H. Lehmer (1905–1991) was able to establish that $2^{257} - 1$ was *not* prime. Subsequently, the Mersenne numbers up to $2^{521} - 1$ were found to be not prime. A number as huge as $2^{521} - 1$, Newman realized, was probably beyond the Baby's scope; his objective, however, was less to make a discovery than to assess the computer's capacities. Accordingly, he set the baby to the task of testing Mersenne primes, using Lucas's method, which required it first to divide the numbers in question into blocks of 40 digits each and then to program the necessary carrying. In the end, though it found no new primes, the Baby was able to verify both Lucas's and Lehmer's findings—no mean feat, and a good indication of its potential.*

Operating the Manchester machine wasn't easy. Among other tasks, the operator had frequently to run from the machine room to the tape room upstairs, where the engineer would, on her instructions, switch the writing current on and then off again. A great amount of physical energy had to be expended, and there was vast room for error. "As every vehicle that drove past was a potential source of spurious digits," Cicely Popplewell later recalled, "it usually took many attempts to get a tape in—each attempt needing another trip up to the tape room." Indeed, the members of the Manchester team were soon so lost in the technical complexities of actually getting the machine to do its job that when news of their

*Julia Robinson later proved that $2_{521} - 1$ was, in fact, prime.

The Manchester Computer in 1955. (© Hulton-Deutsch Collection/CORBIS)

research reached the press, they were ill-prepared to deal with the consequences. And as it happened, the 1948 publication of a book called *Cybernetics*, by the American Norbert Wiener (1894–1964), had started a chain of events that cast upon the Manchester project an unwanted spotlight.

What happened was this: Wiener, who admired Turing, made a special trip to visit him in the spring of 1947 in order to discuss the future of intelligent machines. Wiener's writings were much more sensationalistic than Turing's, in addition to which he was something of a futurist manqué, inclined to play up (for instance) the similarity between nerves and electrical circuits and to prophesy scenarios in which robots working at factories render their human counterparts redundant.

Word of Wiener's ideas and his visit soon reached the ears of Sir Geoffrey Jefferson (1886–1961), the chair of the Department of Neurosurgery at Manchester University and an early advocate of the frontal lobotomy. Jefferson was due to give the Lister Oration at Manchester on June 9, 1949, and chose as his topic "The Mind of Mechanical Man." In effect, the purpose of the speech was to expose and debunk the Manchester computer project, while hymning the innate superiority of the human soul to anything mechanical or man-made:

> Not until a machine can write a sonnet or compose a concerto because of thoughts and emotions felt, and not by the chance fall of symbols, could we agree that machine equals brain—that is, not only write it but know that it had written it. No mechanism could feel (and not merely artificially signal, an easy contrivance) pleasure at its success, grief when its valves fuse, be warmed by flattery, be made miserable by its mistakes, be charmed by sex, be angry or miserable when it cannot get what it wants.

In his report for the NPL, Turing had also addressed, in a rather tongue-in-cheek way, the claim that even if provided with a method of locomotion and sense organs, a machine would still be incapable of enjoying much of what human beings enjoyed. For Turing, however, this was of no consequence: as he later put it, the ability to enjoy strawberries and cream was not a prerequisite for intelligence. Jefferson, on the other hand, brandished the machine's supposed lack of consciousness as evidence of its ultimate stupidity. Summarizing his oration the next day, the *Times* of London paraphrased him as saying that unless a machine could "create concepts and find for itself suitable words in which to express them . . . it would be no cleverer than a parrot"; the paper also reported that Jefferson "feared a great many airy theories would arise to tempt them against their better judgment, but he forecast that the day would never dawn when the gracious rooms of the Royal Society would be converted into garages to house the new fellows."

This was clearly meant as a slight to Newman, whose project the Royal Society had funded, and a day later the newspaper followed up with an article on Newman's "mechanical brain," noting that the "mechanical mind" had "just completed, in a matter of weeks, a problem, the nature of which is not disclosed, which was started in the seventeenth century and is only just being calculated by human beings." The machine was described as being "composed of racks of electrical apparatus consisting of a mass of untidy wires, valves, chassis, and display tubes. When in action, the cathode ray becomes a pattern of dots which shows what information is in the machine. There is a close analogy between its structure and that of the human brain." The article also included an interview with Turing, who said of the machine,

> This is only a foretaste of what is to come, and only the shadow of what is going to be. We have to have some expe-

> rience with the machine before we really know its capabilities. It may take years before we settle down to the new possibilities, but I do not see why it should not enter any of the fields normally covered by the human intellect and eventually compete on equal terms.
>
> I do not think you can even draw the line about sonnets, though the comparison is perhaps a little bit unfair because a sonnet written by a machine will be better appreciated by another machine!

There was no reason to assume, in other words, that even poetry (Jefferson had ended his oration by quoting from *Hamlet*) should be the exclusive province of the human imagination. (A relative who read the article told Mrs. Turing, "Isn't that just like Alan?") Yet what is more striking than Turing's willingness to attribute to a machine the capacity for writing and understanding verse is his suggestion that machines might speak between themselves a language no less meaningful for its exclusion of human beings. It was as if what offended Turing, even more than Jefferson's avidity to shut down avenues of exploration, was his hawking of "humanist" values for the explicit purpose of denying a whole class of beings the right to a mental existence. Likewise homosexual men, for decades, had been erased from history—and more specifically, from the history of human eros to which Jefferson alluded by mentioning "the charm of sex." In any case, Turing told the *Times*, "The university was really interested in the investigation of machines for their own sake." It was as if, by this point, he was becoming sick of the human.

As for Newman, he gave his own reply to the *Times* in the form of a letter published on June 14, in which he attempted to summarize some of the science behind the prototype Manchester machine and also clear the air regarding "the rather mysterious description" that the newspaper had given of the problem dating back to the seventeenth century. Testing out the Mersenne primes, he explained, was exactly the sort of pure mathematical exercise at which Newman hoped his machine would excel. Indeed, the earnestness with which he attempted to make the experiment comprehensible to the *Times*'s readers provided clear evidence as to just how far apart the perspective of the Manchester laboratory was from the one that informed Jefferson's oration. Nonetheless, the letters column of the *Times* continued, for a few days, to offer evidence that perhaps Turing and Newman were underestimating the hostility that their research had the potential to provoke. England was as disinclined to accept machine nature as human nature, if Illtyd Trethowan of Downside Abbey, Bath, was to be believed; in a letter to the *Times* dated June 13, he expressed his hope that "responsible scientists will be quick to dissociate themselves" from Newman's program. "But we must all take warning from it. Even our dialectical materialists would feel necessitated to guard themselves, like Butler's Erewhonians, against the possible hostility of the machines."*

*Turing might have been thinking of Trethowan when at the end of his 1951 Manchester lecture he remarked, "There would be plenty to do in trying, say, to keep one's intelligence up to the standard set by the machines, for it seems probable that once the machine thinking method had started, it would not take long to outstrip our feeble powers. There would be no question of the machines dying, and they would be able to converse with each other to sharpen their wits. At some stage therefore we should have to expect the machines to take control, in the way that is mentioned in Samuel Butler's *Erewhon*."

As for Jefferson's remark that unless a machine could "create concepts and find for itself suitable words in which to express them . . . it would be no cleverer than a parrot," it provoked a spirited defense of the bird in the paper's editorial pages that brought an end to the brouhaha, with the author facetiously complaining that

> those who have never loved a parrot can hardly appreciate the vehemence of the emotions aroused by these thoughtless words in the breasts of those who have made of this sagacious bird a close and (as far as can be ascertained) devoted companion. . . . Parrots can make things devilish unpleasant if they take a dislike to you, and it would be a prudent as well as a courteous gesture if Professor Jefferson withdrew an observation which has ruffled so many and so well-loved feathers.

So far as the *Times* was concerned, the call for an apology to the parrot (but not to the scientists) brought the matter to a close. But Turing did not forget what Jefferson had said. If anything, the exchange in the newspaper's pages only strengthened his interest in machine intelligence. He would soon strike back—and even the parrot would make another appearance.

2.

"Computing Machinery and Intelligence," Alan Turing's most famous and in many ways most perverse paper, appeared in *Mind* in October 1950. Whereas in the NPL report he started with likely objections, here he saved the list of potential objections to computer intelligence for later, and began instead with a clear statement of his intent. "I propose to consider the question, 'Can machines think?' This should begin with definitions of the meaning of the terms 'machine' and 'think.' " But if these meanings "are to be found by examining how they are commonly used it is difficult to escape the conclusion that the meaning and the answer to the question, 'Can machines think?' is to be sought in a statistical survey such as a Gallup poll." Such an idea, in Turing's view, was "absurd."

Instead of offering definitions, Turing recast his question by proposing what he called the imitation game. It would later become known as the Turing test, much as the *a*-machine of "Computable Numbers" has come to be called a Turing machine. The game, as he explains it,

> is played with three people, a man (A), a woman (B), and an interrogator (C) who may be of either sex. The interrogator stays in a room apart from the other two. The object of the game for the interrogator is to determine which of the other two is the man and which is the woman. He knows them by labels X and Y, and at the end of the game he says either "X is A and Y is B" or "X is B and Y is A." The interrogator is allowed to put questions to A and B thus:
>
> C: Will X please tell me the length of his or her hair?
>
> Now suppose X is actually A, then A must answer. It is A's object in the game to try and cause C to make the wrong identification. His answer might therefore be
>
> "My hair is shingled, and the longest strands are about nine inches long."
>
> In order that tones of voice may not help the interrogator the answers should be written, or better still, type-

> written. The ideal arrangement is to have a teleprinter communicating between the two rooms. Alternatively the question and answers can be repeated by an intermediary. The object of the game for the third player (B) is to help the interrogator. The best strategy for her is probably to give truthful answers. She can add such things as "I am the woman, don't listen to him!" to her answers, but it will avail nothing as the man can make similar remarks.
>
> We now ask the question, "What will happen when a machine takes the part of A in this game?" Will the interrogator decide wrongly as often when the game is played like this as he does when the game is played between a man and a woman? These questions replace our original, "Can machines think?"

Turing's proof, in "Computable Numbers," that the *Entscheidungsproblem* was insoluble relied on the ingenious substitution of a complicated question—can a machine decide whether a statement is provable?—with a simpler one: does a certain machine ever print a 0? Along the same lines, in "Computing Machinery and Intelligence" he argued that the complicated question "Can machines think?" could be substituted with the simpler question "Can a machine win the imitation game?" The two, in Turing's view, were identical, because behavior, as he saw it, *was* identity. And yet to apply such a mathematically precise notion of identity to the murky matter of what "human" meant was to invite all sorts of objections—and problems.

For example, the ambiguity of Turing's query "What will happen when the machine takes the part of A?" has occasioned much debate. Does Turing mean to say that instead of being played between a man and a woman, the game should be played between a man and a machine? The rest of the paper would seem to bear out this interpretation. Yet a literal reading of the paragraph suggests a different meaning: that the game should now be played between a man *and a computer pretending to be a man pretending to be a woman.* Hodges shows little patience for this reading, going so far as to argue that "Turing's gender-guessing analogy detracts from his own argument. . . ." After all, as he points out, the section that follows the troublesome paragraph is entirely concerned with the ways in which a machine might trick an interrogator into believing that he (or she) was talking to a human being—male *or* female:

> The new problem has the advantage of drawing a fairly sharp line between the physical and the intellectual capacities of a man. No engineer or chemist claims to be able to produce a material which is indistinguishable from the human skin.* It is possible that at some time this might be done, but even supposing this invention available we should feel there was little point in trying to make a "thinking machine" more human by dressing it up in such artificial flesh. The form in which we have set the problem reflects this fact in the condition which prevents the interrogator from seeing or touching the other competitors, or hearing their voices. Some other advantages of the proposed criterion may be shown up by specimen questions and answers. Thus:
>
> Q: Please write me a sonnet on the subject of the Forth Bridge.

*For an interesting analysis of skin imagery—of which there is a lot—in Turing's paper, see Jean Lassègue, "What Kind of Turing Test Did Turing Have in Mind?" *http://tekhnema.free.fr/3Lasseguearticle.htm.*

A: Count me out on this one. I never could write poetry.

Q: Add 34957 to 76764.

A: (Pause about 30 seconds and then give as answer) 105621.

Q: Do you play chess?

A: Yes.

Q: I have K at my K1, and no other piece. You have only K at K6 and R at R1. What do you play?

A: (After a pause of 15 seconds) R-R8 mate.

Hodges is correct to observe that gender plays no role in the answers given here (including the incorrect addition). And yet to ignore the subtext that Turing's ambiguity exposes is also to ignore the palpable tone of sexual anxiety that runs all through the paper. For instance, just a few paragraphs after the dialogue quoted above, Turing writes, "It might be urged that when playing the 'imitation game' the best strategy for the machine may possibly be something other than imitation of the behaviour of a man. . . . In any case there is no intention to investigate here the theory of the game, and it will be assumed that the best strategy is to try to provide answers that would naturally be given by a man."* Trying to provide "answers that would naturally be given by a man" would, of course, also be the best strategy for a homosexual to adopt when trying to persuade an interrogator that he is straight; in this alternative version of the imitation game, he would talk about cricket and

*Turing used similar language during a 1952 BBC roundtable discussion, in which, as an example of the sort of question to use in the imitation game, he proposed the following: "I put it to you that you are only pretending to be a man." In such a case, "the machine would be permitted all sorts of tricks so as to appear more manlike. . . ."

describe the woman he would like to marry. And though the parallel may be accidental—"a man," after all, could as easily mean "a human being" as "a male human being"—Turing's use of the word "naturally" suggests a more heightened awareness of the idea of the "natural" than the situation calls for. Not surprisingly, arguments concerning the naturalness or unnaturalness of homosexuality ran through both antihomosexual diatribes and apologies for homosexuality written in the period, with Oscar Wilde's championing of the artificial often brandished as an ironic defense of "unnatural" love.*

Turing's preoccupation with gender recurs several more times during the course of the paper. In section 3, a discussion of exactly what defines a "machine" concludes with this rather bizarre proviso that

> we wish to exclude from the machines men born in the usual manner.
>
> It is difficult to frame the discussion so as to satisfy [this condition]. One might for instance insist that the team of engineers should be all of one sex, but this would not really be satisfactory, for it is probably possible to rear a complete individual from a single cell of the skin (say) of a man. To do so would be a feat of biological technique deserving of the very highest praise, but we would not be inclined to regard it as a case of "constructing a thinking machine."

Is the point here that the team of engineers—all of "one sex"—might be able to join together and in a sort of orgy of

*In *Maurice* the hero asks Alec, "Scudder, why do you think it's 'natural' to care both for men and women? You wrote so in your letter. It isn't natural for me. I have really got to think that 'natural' only means oneself."

cloning create a human child? The fantasy is peculiar, using science as a framework for imagining a means by which men without women could generate progeny. Of course, Turing too longed to produce a child of his own—a computer child. It is therefore not surprising that in the paper he soon returns to the metaphor of child rearing and education, employing a "domestic analogy" to describe the ways in which a machine might be taught to obey "not fresh instructions on each repetition, but the same ones over and over again":

> Suppose Mother wants Tommy to call at the cobbler's every morning on his way to school to see if her shoes are done, she can ask him afresh every morning. Alternatively she can stick up a notice once and for all in the hall which he will see when he leaves for school and which tells him to call for the shoes, and also to destroy the notice when he comes back if he has the shoes with him.

"Tommy" is here the computer, the offspring of a group of engineers who have eschewed cloning in favor of other styles of cooperation—perhaps the sort of cooperation in scientific experiment that Turing so cherished in his friendship with Christopher Morcom. More importantly, Tommy is a digital computer, and in Turing's estimation only a digital computer—a universal machine—has a shot at ever winning the imitation game.

> I believe that in about fifty years' time* it will be possible to programme computers, with a storage capacity of about

*By 1952, when he was interviewed on the BBC, the estimate had gone up to at least a hundred years.

> 10^9, to make them play the imitation game so well that an average interrogator will not have more than 70 per cent chance of making the right identification after five minutes of questioning.

The Manchester "Baby" is clearly growing up.

3.

By this point, then, a subtle but distinct strain of anxiety concerning gender, sexual imitation, and even homosexual procreation has come to assert itself within Turing's "official" argument about machine intelligence. But where does it come from? The answer can be traced back to Sir Geoffrey Jefferson's Lister Oration, the slightly masculinist tone of which Turing ridicules in the paper, even as he rebuts Jefferson's "humanist" stance. This is especially evident near the middle of "Computing Machinery and Intelligence," where Turing takes up once again the strategy of listing—and then refuting—the objections that might be raised to the possibility of a thinking machine. Although Professor Jefferson does not appear by name until the fourth objection—"The Argument from Consciousness"—his spirit is invoked, and mocked, from the very start.

For instance, in his refutation of the first objection—"the theological objection" that "God has given an immortal soul to every man and woman, but not to any other animal or to machines"—Turing questions the implicit superiority of mankind that provided the basis for Jefferson's diatribe, noting, "I should find the argument more convincing if animals were classed with men, for there is a greater difference, to my mind, between the typical animate and the inanimate than there is between man and the other animals." Likewise, how are Christians to contend with "the Moslem view that women have no souls"? By invoking the rights not just of women but of animals, Turing allies himself (and his computer) with all the other populations that have suffered at the hand of religions that take the superiority of man (in one case) and mankind (in the other) for granted. Against this he posits his own rather odd theology, which, needless to say, blesses machines, by equating their construction with procreation: "In attempting to construct such machines we should not be irreverently usurping His power of creating souls, any more than we are in the procreation of children: rather we are, in either case, instruments of His will providing mansions for the souls that He creates."

The assumption of mankind's innate superiority is challenged even more boldly in Turing's retort to the second objection, which he calls the "Heads in the Sand" objection and sums up as follows: "The consequences of machines thinking would be too dreadful. Let us hope and believe that they cannot do so." This, of course, was the very posture to which some of Norbert Wiener's writings inadvertently appealed, and in responding to it, Turing also responds to Jefferson, noting that the feeling that mankind is "*necessarily* superior" to the rest of creation "is likely to be quite strong in intellectual people, since they value the power of thinking more highly than others, and are more inclined to base their belief in the superiority of Man on this power." With his allusions to Shakespeare, Jefferson is exemplary of these "intellectual people" whose tendency to exalt their own species Turing shows so little patience for. It is a point he returns to in his answer to the third objection, the "mathematical objection," which is essentially the argument (paraphrased in the NPL report) that his own resolution of the *Entscheidungsproblem*, in conjunction with Gödel's findings, proves "that there are certain things that . . . a machine cannot do." Turing was obviously made uncomfortable by the possibility that his solution to the *Entscheidungsproblem* might be employed in an attack on the machine that the *Entscheidungsproblem* propelled him to create. In responding to it here, however, he focuses squarely on the psychology of what might be called the natural "superiority complex" of human beings (especially intellectuals), observing shrewdly that when a machine gives a wrong answer to

> the appropriate critical question . . . this gives us a certain feeling of superiority. Is this feeling illusory? It is no doubt quite genuine, but I do not think too much importance should be attached to it. We too often give wrong answers to questions ourselves to be justified in being very pleased at such evidence of fallibility on the part of machines. Further, our superiority can only be felt on such an occasion in relation to the one machine over which we have scored our petty triumph. There would be no question of triumphing simultaneously over *all* machines.

Here Turing seems to be amusing himself, in a rather quiet way, by alluding to Mr. Illtyd Trethowan's anxiety about "the possible hostility of the machines," over *all* of whom we can never hope to triumph. More importantly, this rebuttal gives him the chance to repeat one of his key points—that fallibility is a key ingredient in intelligence.

It is in his refutation of objection 4—"the Argument from Consciousness"—that Turing takes direct aim at Jefferson, whom he begins by quoting and at whom he hurls one of his most memorable and witty retorts:

> This argument appears to be a denial of the validity of our test. According to the most extreme form of this view the only way by which one could be sure that a machine thinks is to *be* the machine and to feel oneself thinking. . . . Likewise according to this view the only way to know that a *man* thinks is to be that particular man. It is in fact the solipsist point of view. It may be the most logical view to hold but it makes communication of ideas difficult. A is liable to believe "A thinks but B does not" whilst B believes "B thinks but A does not." Instead of arguing continually over this point it is usual to have the polite convention that everyone thinks.

Rather cleverly, Turing writes that he is "sure that Professor Jefferson does not wish to adopt the extreme and solipsist point of view." He then compares his own imitation game with a game called *viva voce*, the purpose of which is "to discover whether some one really understands something or has 'learned it parrot fashion.'" Notably, the exemplary *viva voce* that Turing cites is replete with literary references, with the questioner first asking his subject about Shakespeare, then veering into Dickens. The point is that the imitation game *also* determines whether someone has learned something "parrot fashion"; it differs from *viva voce* only in that the person being tested is a machine. Nor is it a coincidence that literature plays such a prominent role in this *viva voce*, the orchestrator of which is presumably a self-proclaimed intellectual such as Jefferson. And surely any self-respecting intellectual would rather abandon the argument from consciousness "than be forced into the solipsist position."

Having got rid of Jefferson—at least in name—Turing next addresses a whole class of objections that he calls "Arguments from Various Disabilities," and which he defines as taking the form "I grant you that you can make machines do all the things you have mentioned but you will never be able to make one to do X." He then offers a rather tongue-in-cheek "selection":

> Be kind, resourceful, beautiful, friendly; have initiative, have a sense of humour, tell right from wrong, make mistakes; fall in love, enjoy strawberries and cream; make some one fall in love with it, learn from experience; use words properly, be the subject of its own thought; have as much diversity of behaviour as a man, do something really new.

As Turing notes, "no support is usually offered for these statements," most of which are

> founded on the principle of scientific induction. . . . The works and customs of mankind do not seem to be very suitable material to which to apply scientific induction. A very large part of space-time must be investigated, if reliable results are to be obtained. Otherwise we may (as most English children do) decide that everybody speaks English, and that it is silly to learn French.

Turing's repudiation of scientific induction, however, is more than just a dig at the insularity and closed-mindedness of England. His purpose is actually much larger: to call attention to the infinite regress into which we are likely to fall if we attempt to use disabilities (such as, say, the inability, on the part of a man, to feel attraction to a woman) as determining factors in defining intelligence. Nor is the question of homosexuality far from Turing's mind, as the refinement that he offers in the next paragraph attests:

> There are, however, special remarks to be made about many of the disabilities that have been mentioned. The inability to enjoy strawberries and cream may have struck the reader as frivolous. Possibly a machine might be made to enjoy this delicious dish, but any attempt to make one do so would be idiotic. What is important about this disability is that it contributes to some of the other disabilities, e.g. to the difficulty of the same kind of friendliness occurring between man and machine as between white man and white man, or between black man and black man.

To the brew of gender and sexuality, then, race is added, as "strawberries and cream" (earlier bookended between the ability to fall in love and the ability to make someone fall in love) becomes a code word for tastes that Turing prefers not to name. In many ways the passage recalls the rather campy bathhouse scene in the 1960 film *Spartacus*, in which a dialogue about other "delicious dishes" encodes a subtle erotic bargaining between Crassus (Laurence Olivier) and his slave Antoninus (Tony Curtis).

> *Crassus:* Do you eat oysters?
> *Antoninus:* When I have them, master.
> *Crassus:* Do you eat snails?

Antoninus: No, master.

Crassus: Do you consider the eating of oysters to be moral, and the eating of snails to be immoral?

Antoninus: No, master.

Crassus: Of course not. It's all a matter of taste.

Antoninus: Yes, master.

Crassus: And taste is not the same as appetite and therefore not a question of morals, is it?

Antoninus: It could be argued so, master.

Crassus: Um, that'll do. My robe, Antoninus. Ah, my taste . . . includes both oysters and snails.

In this exchange Crassus, too, is engaging in a kind of imitation game, the purpose of which is to assess whether it would or would not be a good idea to offer Antoninus (who prefers oysters) some of his snails. Antoninus, at the same time, recognizes the advantage, at least on occasion, of giving the "wrong" answer ("No, master")—just as a machine would have to if it were to have a chance of winning the game:

> The claim that "machines cannot make mistakes" seems a curious one. . . . I think this criticism can be explained in terms of the imitation game. It is claimed that the interrogator could distinguish the machine from the man simply by setting them a number of problems in arithmetic. The machine would be unmasked because of its deadly accuracy. The reply to this is simple. The machine (programmed for playing the game) would not attempt to give the *right* answers to the arithmetic problems. It would deliberately introduce mistakes in a manner calculated to confuse the interrogator.

"Errors of functioning," then, must be kept distinct from "errors of conclusion." Nor should it be assumed that machines are not capable of deception. On the contrary, the criticism "that a machine cannot have much diversity of behaviour is just a way of saying that it cannot have much storage capacity."

Turing wraps up his catalog of possible objections to the thinking machine with four rather curious examples. The first, which he calls "Lady Lovelace's Objection" (in reference to Byron's daughter and Babbage's muse), is that computers are incapable of "originating" anything. Instead (and here Turing quotes Lady Lovelace), "a computer can do *whatever we know how to order it* to perform." But, as Turing points out, in actual practice, machines surprise human beings all the time. Turing then rebuts the "Argument from Continuity in the Nervous System"—although it is true that a discrete-state machine cannot mimic the behavior of the nervous system, "if we adhere to the conditions of the imitation game, the interrogator will not be able to take any advantage of this difference"—and assesses the "Argument from Informality of Behavior": "If each man had a definite set of rules of conduct by which he regulated life he would be no better than a machine. But there are no such rules, so men cannot be machines." This objection Turing answers, first, by distinguishing "rules of conduct" from the "laws of behavior" by which machines are presumably regulated, then by pointing out that "we cannot so easily convince ourselves of the absence of complete laws of behaviour as of complete rules of conduct." By way of example, he describes another experiment:

> I have set up on the Manchester computer a small programme using only 1000 units of storage, whereby the machine supplied with one sixteen figure number replies with another within two seconds. I would defy anyone to learn from these replies sufficient about the program to be able to predict any replies to untried values.

The last—and most peculiar—objection that Turing takes on is the argument "from extra-sensory perception," which he prefaces with a surprisingly credulous description of telepathy, clairvoyance, precognition, and psychokinesis. Of these he remarks, "Unfortunately the statistical evidence, at least for telepathy, is overwhelming. It is very difficult to rearrange one's ideas so as to fit these new facts in." Without giving a source for this "overwhelming" evidence, Turing goes on to give the "strong" argument from ESP against a machine's winning the imitation game:

> Let us play the imitation game, using as witnesses a man who is good as a telepathic receiver, and a digital computer. The interrogator can ask such questions as "What suit does the card in my right hand belong to?" the man by telepathy or clairvoyance gives the right answer 130 times out of 400 cards. The machine can only guess at random, and perhaps gets 104 right, so the interrogator makes the right identification.

For Turing, the scenario as described opens up the "interesting possibility" of equipping the digital computer in question with a random number generator.

> Then it will be natural to use this to decide what answer to give. But then the random number generator will be subject to the psycho-kinetic powers of the interrogator. Perhaps this psycho-kinesis might cause the machine to guess right more often than would be expected on a probability calculation, so that the interrogator might still be unable to make the right identification. On the other hand, he might be able to guess right without any questioning, by clairvoyance. With E. S. P. anything may happen.

Rather than offering a refutation of this argument, Turing says only that perhaps the best solution would be to put the competitors into a "telepathy-proof room"—whatever that means. One wonders what the editors of that august scientific publication *Mind* made of this bizarre appeal to a pseudoscience as baseless, if not as pernicious, as the one on the altar of which Turing would soon be laid out, as a kind of experiment. For how could they know that years before Turing had loved a boy named Christopher Morcom, with whose spirit he had been determined to remain connected even after death?*

"Computing Machinery and Intelligence" concludes with a meditation on teaching and learning that reiterates much of the technique prescribed in "Intelligent Machinery." Here, however, Turing adds the proviso that his system of punishments and rewards does not "presuppose any feelings on the part of the machine." Moving a bit away from the rigorously behaviorist ethos that animated "Intelligent Machinery," he

*See Lassègue, "What Kind of Turing Test Did Turing Have in Mind?," for an interesting discussion of the role Christopher Morcom might have played—even subliminally—in the paper.

also reminds his readers that "the use of punishments and rewards can at best be a part of the teaching process. . . . By the time a child has learnt to repeat 'Casabianca' he would probably feel very sore indeed, if the text could only be discovered by a 'Twenty Questions' technique, every 'NO' taking the form of a blow." Less emotional techniques need to be employed as well, especially when the objective is to teach the machine to obey orders in a symbolic language.

Probably the biggest shift from "Intelligent Machinery," however, is that here Turing elects to anthropomorphize his child-machine to a much greater degree than in the earlier paper, putting more emphasis on its childishness than on its machinishness. For example, near the end of the paper, he asks, "Instead of trying to produce a programme to simulate the adult mind, why not rather try to produce one which simulates the child's? . . . Presumably the child-brain is something like a note-book as one buys it from the stationer's. Rather little mechanism, and lots of blank sheets." But because this notebook mind is contained within a machine body, a slightly different teaching process has to be applied to it than would be to the "normal" child:

> It will not, for instance, be provided with legs, so that it could not be asked to go out and fill the coal scuttle. Possibly it might not have eyes. But however well these deficiencies might be overcome by clever engineering, one could not send the creature to school without the other children making excessive fun of it. It must be given some tuition. We need not be too concerned about the legs, eyes, etc. The example of Miss *Helen Keller* shows that education can take place provided that communication in both direc-

> tions between teacher and pupil can take place by some means or other.

One thinks of Turing as a boy, "watching the daisies grow." Does he feel some sense of identification with Helen Keller, provided (as Turing was not) with an education to suit her particular disabilities? Certainly

> the imperatives that can be obeyed by a machine that has no limbs are bound to be of a rather intellectual character. . . . For at each stage when one is using a logical system, there is a very large number of alternative steps, any of which one is permitted to apply, so far as obedience to the rules of the logical system is concerned. These choices make the difference between a brilliant and a footling reasoner, not the difference between a sound and a fallacious one.

And the ability to reason is, finally, the ultimate evidence of intelligence. If it is to be attained, however, flexibility is essential, even if "the rules which get changed in the learning process are of a rather less pretentious kind, claiming only an ephemeral validity. The reader may draw a parallel with the Constitution of the United States."

In the end, Turing believes, the goal should be to do exactly what alarms Jefferson: to construct machines that "will eventually compete with men in all purely intellectual fields." Perhaps the best way to start would be to teach the machine some "very abstract activity," such as how to play chess; or perhaps it would make more sense to provide it with "the best sense organs that money can buy, and then teach it to understand and speak English." In either case, the final note that

Turing sounds in "Computing Machinery and Intelligence" combines triumph with a certain detached self-assurance. For Turing, thinking machines are inevitable, whether we like them or not. It is as if his faith in future tolerance had once again bolstered him against the very real threat of present injustice.

4.

The years Turing spent working with the Manchester computer were marked by an increasing isolation from other people, as he became less and less interested in the computer itself and more and more involved in the experiments he was using it for. Not that he only did experiments: he also wrote a programmer's handbook in which he urged potential users of the Manchester machine to employ an almost literary sensibility in designing programs. Most of his time, though, he devoted to the application of the machine to such pure mathematical problems as constructing a new proof for the word problem for semigroups, and to working with permutation theory, which had played an important role in his code breaking at Bletchley. His colleague Christopher Strachey also taught the machine to sing "God Save the King."

Probably the experiment that meant the most to Turing, however, was the one with which he had the least success. For years he had remained fascinated by the Riemann hypothesis, which for some reason he had convinced himself had to be false. True, the machine he had tried to build with Donald MacPhail at Cambridge had ended up on the scrap heap. Yet he had never forgotten his ambition of beating Titchmarsh's record for the calculation of zeros, and still hoped he might one day be able to find a zero *off* the critical line. Toward that end, in 1943 he had published a paper entitled "A Method for the Calculation of the Zeta-function" in the *Proceedings of the Mathematical Society*. Titchmarsh, using hand methods, had shown that all the zeros up to t = 1,468 were on the critical line. Now Turing put his own method to the test. In 1953 he designed a program by means of which the Manchester computer could calculate zeta zeros using its complex base-32 code, and by means of that program he proved the validity of the Riemann hypothesis as far as t = 1,540—72 more zeros than Titchmarsh had found—before the machine broke down.

It was, as Turing ruefully noted, "a negligible advance."

8

Pryce's Buoy

1.

In the spring of 1951 Alan Turing was elected a member of the Royal Society. Among the congratulatory notes he received was one from his old antagonist Sir Geoffrey Jefferson, who wrote, "I am so glad; and I sincerely trust that all your valves are glowing with satisfaction, and signalling messages that seem to you to mean pleasure and pride! (but don't be deceived!)."

As it happened, Turing and Jefferson were destined to tangle once more. The occasion was a roundtable discussion of machine intelligence broadcast on the BBC Third Programme on January 14, 1952, in which the other participants were Max Newman and Turing's old Cambridge friend Richard Braithwaite, one of the two mathematicians who had long ago asked for offprints of "Computable Numbers." Braithwaite acted as moderator, and while the conversation did little to advance the cause of the thinking machine, at the very least it gave the speakers a chance to refine and clarify some of their positions. As always, Jefferson insisted that it was the "high emotional content of mental processes in the human being that makes him quite different from a machine," while Newman—ever the pragmatist—struggled valiantly to keep the focus on what existing machines could actually do, as opposed to what future machines *might* do. Much airtime was devoted to gratuitous speculation; Braithwaite wanted to know whether it would be legitimate to ask the machine what it had had for breakfast, while Jefferson insisted that in creating a model of actual thinking, "the intervention of extraneous factors, like the worries of having to make one's living, or pay one's taxes, or get food that one likes" could not be "missed out." Newman, in turn, emphasized pure mathematics and quoted the sculptor Henry Moore as saying, "When the work is more than an exercise, inexplicable jumps occur. This is where the imagination comes in."

As for Turing, his answers to the questions put to him were tinged with weariness, perhaps at having to defend his ideas for the thousandth time against the same objections. Once again, he outlined the mathematics of machine learning. Once again, he asserted that intelligence was not the same thing as infallibility. Once again, he reminded his listeners that sometimes "a computing machine does something rather weird that we hadn't expected." He was not able to get much support from Braithwaite, who had a habit of punctuating the discussion with inane observations that only made Turing's job harder; at one point, for instance, Braithwaite wondered whether it would be "necessary for the machine to be able to throw fits or tantrums"—a rather stupefying query that Turing elided by saying that he would be "more interested in curbing such displays than in encouraging them."

Not surprisingly, Jefferson got the last word. A remark to

the effect that he would not be willing to believe a computing machine could think "until he saw it touch the leg of a lady computing machine" was cut from the broadcast; still, he brought things to a jaunty enough conclusion to diminish Turing's seriousness, averring, "That old slow coach, man, is the one with the ideas—or so I think. It would be fun some day, Turing, to listen to a discussion, say on the Fourth Programme, between two machines on why human beings think that they think!" The transcript does not say whether or not Turing laughed in reply.

All told, Turing gave a lackluster performance. Yet if on the BBC he failed to invest his argument for machine intelligence with the passion that had animated his papers on the topic, it was only in part because he was tired of being put on the defensive. Earlier, he had left off his cryptanalysis work to undertake the Delilah speech encipherment machine. Then he had abandoned Delilah to build the ACE at Teddington—a project from which he had in turn drifted away as the theory of artificial intelligence consumed his imagination. Finally, in the early fifties, he was moving away from machines altogether. As Mrs. Turing noted in her memoir, since his childhood her son had been fascinated by biology. At Cambridge he had discussed the similarities between machine circuitry and the brain with Peter Matthews; indeed, all his papers on computer intelligence mention the prospect of building a machine on the model of the brain. Now he took up this intriguing notion from the other side.

The question that preoccupied him was basic: could mathematical models be constructed for the processes of biological growth, or *morphogenesis*, as it was more technically termed? Previously Turing had focused on exploring whether mechanical systems could be designed that mirrored the process of human thought. Now he wanted to know whether mathematical theory might provide a basis for investigating physiology. Armed with differential equations, he was marching into Jefferson's territory—and with the same boldness that marked Jefferson's unwelcome invasion of the Manchester laboratory with its "late lavatorial" tiles. Not that Turing was inclined to play to the gallery, as Jefferson did; instead, he made sure that his approach was rigorously, even bewilderingly mathematical.*

As a boy, Turing had been drawn to watching the daisies grow. Now he wanted to figure out exactly what he had seen. Although the work had nothing to do with computer design, in many ways it represented a more fitting culmination to his intellectual career than even building the ACE would have. For in the end Turing was more a child of Hardy than of von Neumann, which meant that understanding the world mattered more to him than changing it.

Disquisiting on the BBC, he had insisted that he was "not interested in the fact that the brain has the consistency of cold porridge. We don't want to say 'This machine's quite hard, so it isn't a brain, and so it can't think.'" Still, hard machines had obsessed him for years, as he had taken up the cause of the man-made with an avidity to match Oscar Wilde's. And now, suddenly, here he was writing about "natural" processes. More to the point, he was living, for the first time, something like

*As Newman explained, "Turing had arrived at differential equations of the form $\nabla^2 x_i$ for n different morphogens in continuous tissue; where f_i is the reaction function given the rate of growth of X_i, and $\frac{\delta x_i}{\delta t} = f_i(x_i, \ldots, x_n) + \mu\nabla^2 x_i$, $(i = 1, \ldots, n)$ is the rate of diffusion of X_i. He also considered the corresponding equations for a set of discrete cells."

the "natural" life that Forster's Maurice disdained: owning a house, going to work each day, even employing a housekeeper, Mrs. Clayton, with whom, his mother wrote,

> he shared many jokes, for he delighted to regale her with tales against himself. There was, for instance, the occasion when, his watch being under repair, he carried a little clock in his pocket. Suddenly in the crowded train to Manchester the alarm went off and everyone in the compartment jumped. On his runs he often forgot to take his door-key, so one was kept hidden near the spout of the garage gutter. One day he knocked it over the spout and it just slipped away into the ground, a fact which he reported to Mrs. Clayton with much relish.

In other words, he was still Turing—just Turing with a fixed address, and friends. These included his neighbors, Mr. and Mrs. Roy Webb (he was also fond of their very young son, Rob); Max Newman and his wife, the novelist Lyn Newman; Robin Gandy; and, perhaps most importantly, Norman Routledge, also of King's, who was gay and in whom Turing could therefore confide about matters he was disinclined to bring up with the others. For his erotic life, if not flourishing exactly, seemed at least to have become a less depressing business than it had been previously. Although his move to Manchester had put something of a damper on his relationship with Neville Johnson, he traveled often to Europe—on one occasion to Norway and on several to France. The Napoleonic Code—which did *not* criminalize sex between men—meant that while "abroad," Englishmen could enjoy a much needed respite from the aura of worry and guilt that still attached itself to homosexual sex in Britain. The Continent gave Turing the chance to enjoy erotic dalliances without fear of repercussions, and in seeking this freedom, even just for a few days at a time, he was typical of his generation.

Not that he limited his explorations to Europe. In Manchester, too, he enjoyed the occasional adventure or affair. One of these began in January 1952, around the same time that the BBC broadcast his debate with Jefferson, when he picked up a nineteen-year-old boy named Arnold Murray outside the Regal Cinema. Like many working-class youths at the time, Arnold was both underfed and more or less penniless; nor was it unheard of for a boy in his position to earn a little extra cash by going behind the arches at the railroad station with an older man. And yet both of them seem to have wanted something more than this. Accordingly Turing took Arnold out to lunch, then invited him to come to his house that weekend. Arnold accepted the invitation, but never showed up. They met again in Manchester the following Monday, at which point Turing proposed that this time Arnold come home with him immediately. Arnold agreed. He visited Turing a second time later that month, for dinner and (apparently) to spend the night. Afterward Turing sent him a penknife.

There was some confusion about money. Not wanting to be treated as a prostitute, Arnold rebuffed Turing's efforts to give him cash. Then Turing discovered money missing from his wallet. Arnold denied having had anything to do with it, but asked for a loan of three pounds to pay off a debt. A few days later he requested a further seven pounds—again, to pay off a debt. They had a brief wrangle when Turing asked to whom the money was owed, but in the end he wrote Arnold a check.

Several days afterward Turing's house was broken into. The thief—or thieves—made off with about fifty pounds worth of his belongings: clothes, some fish knives, a razor, and a compass, among other things. He called the police, and two officers fingerprinted the house. Then—suspicious that Arnold might have been involved—Turing consulted his neighbor's solicitor, who advised him to write Arnold a letter reviving the matter of the money missing from his wallet, reminding him that he owed Turing money, and suggesting that it would be best if they did not see each other again. In response to the letter, Arnold showed up at his house in a rage, threatening to "do his worst." The scene was akin to the one that takes place between Maurice and Alec at the British Museum, with the poor gamekeeper telling the bourgeois stockbroker, "Mr. Hall—you reckernize it wouldn't very well suit you if certain things came out, I suppose." In response, Maurice lashes out: "By God, if you'd spilt on me . . . , I'd have broken you. It might have cost me hundreds, but I've got them, and the police always back my sort against yours." If, however, Turing took it for granted that the police would back his kind against Arnold's, he was dreadfully mistaken.

In *Maurice* the blackmail threat is a prelude to reconciliation: Alec backs off, and the lovers retreat to a hotel. Something similar happened outside of Manchester. First Arnold, having rescinded his threat, came clean, admitting that he had boasted of his affair with Turing to a youth called Harry, who had in turn proposed robbing Turing's house. Arnold had refused to have anything to do with the plan, he said; even so, it was possible that Harry, on his own, had decided to go through with it. The confession led, as in *Maurice*, to a melting of differences, tenderness, and sex.

Arnold promised that he would try to retrieve the stolen goods—and indeed, a few days later, he reported back to Turing that he had already had some success in tracking them down. But by then it was too late. For the police had returned—not to report to Turing on the progress of their investigation, but to tell him that they "knew all about" his affair with Arnold. They had put two and two together, and now, instead of arresting the thief, they arrested his victim. The charge was gross indecency with another male: the same crime of which Oscar Wilde had been convicted, and for which he had been sent down, more than fifty years before.

2.

The little that was left of Alan Turing's life after his arrest was a slow, sad descent into grief and madness. Tried on morals charges, he was "sentenced"—in lieu of prison—to undergo a course of estrogen treatments intended to "cure" him of his homosexuality. The estrogen injections had the effect of chemical castration. Worse, there were humiliating side effects. The lean runner got fat. He grew breasts. Through it all he continued to work, soldiering on with the resilience he had had to learn at Bletchley. When, for instance, Norman Routledge wrote to him in February inquiring about jobs in intelligence, he replied, "I don't think I really do know about jobs, except the one I had during the war, and that certainly did not involve any travelling. I think they do take on conscripts. . . . However I am not at present in a state in which I am able to concentrate well, the reasons explained in next paragraph."

And what restraint, to write such a casual answer to Routledge's query before spilling the bad news!

> I've now got myself into the kind of trouble that I have always considered to be quite a possibility for me, though I have usually [illegible] it at about 10:1 against. I shall shortly be pleading guilty to a charge of sexual offences with a young man. The story of how it all came to be found out is a long and fascinating one, which I shall have to make into a short story one day, but have not time to tell you now. No doubt I shall emerge from it all a different man, but quite who I've not found out.

The letter concludes,

> Glad you enjoyed broadcast. J. certainly was rather disappointing though. I'm rather afraid that the following syllogism may be used by some in the future:
>
> Turing believes machines think
> Turing lies with men
> Therefore machines do not think

It is signed, "yours in distress, Alan."

Now there was no question of his ever again working on government cryptanalysis projects, even though Hugh Alexander had recently approached him about doing just that. He was too much of a security risk. Since the 1951 defection of Guy Burgess, the myth of the homosexual traitor had been gaining momentum, both in the popular press and in the halls of government. Nor had Forster helped matters by writing in his essay "What I Believe" that if pressed to choose between betraying his country and betraying his friend, he hoped he would have the courage to betray his country. If the police now trailed Turing, and even tried to keep him from leaving the country, it was not just to torment him; it was also out of fear that he might decide to betray his country, by delivering the secret information to which he was privy to an enemy agent posing as a friend. It did not matter to them that Turing was genuinely apolitical. He hardly existed for them, and having emasculated him chemically, they now emasculated him morally, by robbing him not just of his freedom to wander but of his freedom to feel. Indeed, it may have been because he felt so emasculated that in a second letter to Norman Routledge—this one from a year later—he chose to employ a self-consciously effeminate tone mostly absent from his earlier correspondence:

> I have a delightful story to tell you of my adventurous life when next we meet. I've had another round with the gendarmes, and it's positively round II to Turing. Half the police of N. England (by one report) were out searching for a supposed boy friend of mine. . . . Perfect virtue and chastity had governed all our proceedings. But the poor sweeties never knew this. A very light kiss beneath a foreign flag, under the influence of drink, was all that had ever occurred. . . . [T]he innocent boy has had rather a raw deal I think. I'll tell you all when we meet in March at Teddington. Being on probation my shining virtue was terrific, and had to be. If I had so much as parked my bicycle on the wrong side of the road there might have been 12 years for me. Of course the police are going to be a bit more nosy, so virtue must continue to shine.

Turing concludes by telling his friend that he might "try to get a job in France." He also confides that he has begun psychoanalysis. The letter is signed, "kisses, Alan."

The psychoanalyst was Dr. Frank M. Greenbaum. Perhaps under his influence, Turing wrote—or at least began to write—the short story he had mentioned to Norman Routledge. Here he cast himself as Alec Pryce (the use of the physicist Maurice Pryce's last name might have been coincidental), a scientist who resembles him in every way except that instead of computer design, his area of expertise is "interplanetary travel." Just as Turing is the father of the Turing machine, Alec is the architect of something called Pryce's buoy—presumably a sort of satellite or spacecraft:

> Alec always felt a glow of pride when this phrase was used. The rather obvious *double-entendre* rather pleased him too. He always liked to parade his homosexuality, and in suitable company Alec would pretend that the word was spelt without the "u." It was quite some time now since he had "had" anyone, in fact it was not since he had met that soldier in Paris last summer. Now that his paper was finished he might justifiably consider that he had earned another gay man, and he knew where he might find one who might be suitable.

Like Turing, Alec has a habit of going on "rather wildly to newspapermen or on the Third Programme." Like Turing, he was also rather untidy, dressing in "an old sportscoat and rather unpressed worsted trousers." One hears the voice of Dr. Greenbaum in the background of the analysis of his sartorial tendencies that comes next:

> He didn't care to wear a suit, preferred the "undergraduate's uniform," which suited his mental age, and encouraged him to believe he was still an attractive youth. This arrested development also showed itself in his work. All men who were not regarded as prospective sexual partners were fellow scholastics to whom Alec had to be actively showing off his intellectual powers.

As the story begins, Alec is Christmas shopping. He is also, presumably, looking out for the "gay man" he feels he has "earned"—and at this point the story's point of view changes rather suddenly to that of "Ron Miller," the stand-in for Arnold. Ron, we learn, "had been out of work for two months, and he'd got no cash. He ought to have had 10s or so for that little job he had helped Ernie over. All he had had to do was to hold the night watchman in conversation for a few minutes whilst the others got on with it. But still it wasn't safe." Ron is also "very hungry and rather cold in the December weather." And, apparently, he is not unwilling to consider bartering sex for cash:

> If he let someone take him under the arches for ten minutes he might get four bob. The men didn't seem to him so keen for it as they were a year ago before the [Ron's?] accident. Of course it wasn't the same as having a girl, nothing like it, but if the chap wasn't too old it wasn't unpleasant. Ernie had said how his chaps would make love to him just as if he were a girl, and say such things! But these were toffs. Ernie with his pansy [illegible] and his pretty-pretty doll's face could get them as easy as [illegible]. Should

> think he liked it quite a lot too, the sorry little swine. Heard [illegible] boast he couldn't do anything with a girl when *she* paid *him!*

Presumably, then, Ron sees himself as being in a different class from Ernie, that "sorry little swine"—just as Arnold saw himself as being in a different class from such friends as, perhaps, the "little swine" Harry. Nonetheless, he is on the lookout and notices that Alec is watching him:

> That chap who was walking round the place had given him quite a look. . . . Here he was coming round again. This time Ron stared back, and Alec followed in his walk and hurried on round the plot again. No doubt now of what he wanted. Didn't seem to have the nerve though. Better give him a little encouragement if he came past again. He was coming too. Ron caught Alec's eye and gave him a half-hearted smile. It was enough though. Alec approached the park seat; Ron made room for him and he sat down. Didn't seem to be very well dressed. What an overcoat! Why wasn't he saying anything? Could he be mistaken? No, he was having a furtive look. . . . [I]f he wasn't careful nothing would come of it.

Ron now asks whether Alec has a cigarette. And, as it happens, Alec does—though this requires some explanation:

> He didn't smoke, because he hadn't quite enough control if he did, and anyway it didn't really agree with him. But he knew that if he "clicked" he would need some cigarettes.
>
> "Doing anything this afternoon?" Alec asked suddenly.

> It was a standard opening. A bit brusque certainly, but he hadn't thought up anything better: anyway the brusqueness tended to prevent misunderstandings. This chap would do well enough. Not a real beauty, but had a certain appeal. Beggars couldn't be choosers. He was shaking his head. "Come and have some lunch with me."

Beggars can't be choosers. What is so sad about this moment is the forcefulness with which the need for Forsterian connection obliterates standards, class, memory (of the idealized Morcom), even self-esteem. And not only for Alec—also (though to a lesser extent) for Ron:

> "Don't mind if I do," said Ron. He didn't go for lah-di-dah ways of talking. It'd come to the same anyway. Bed's bed whatever way you get into it. Alec thought otherwise, and was silent for a couple of minutes as they made their way towards Grenkoff's [?]. He'd got to go through with the lunch at least now. Ron was quite clear about this too. He was sure of the meal. He wasn't sure whether he'd do anything. Perhaps he'd be able to get something without.

In the restaurant, however, Ron is too dazzled by having "the door opened for you by a commissionaire, and to go through first like a girl," to take much notice of Alec. Instead, his attention is "concentrated entirely on the restaurant and its trappings." And this makes Alec happy: "Usually when he went to a restaurant he felt self-conscious, either for being alone, or for not doing the right thing. Ron wouldn't . . ."

And here the story breaks off. We never learn what happens to Ron and Alec. They are left forever on the brink of possibility—perhaps of possible happiness—untouched by the shadow that had swooped down and destroyed the life of their creator.

3.

Mrs. Clayton, his housekeeper, found Alan Turing's dead body in his bed on the morning of June 8, 1954. Nearby was an apple out of which several bites had been taken.

"You will by now have heard of the death of Mr. Alan," she wrote to his mother.

> It was such an awful shock. I just didn't know what to do. So I flew across to Mrs. Gibson's and she rang Police and they wouldn't let me touch or do a thing. & I just couldn't remember your address I had been away for the week end and went up tonight as usual to get his meal. Saw his bedroom light on the lounge curtains not drawn back, milk on step and paper in door. So I thought he'd gone out early & forgot to put his light off. So I went & knocked at his bedroom door. Got no answer so walked in. saw him in bed he must have died during the night. The police have been up here again to night for me to make a statement.

She then added,

> Mr. And Mrs. Gibson saw Mr. Alan out walking Mon. evening he was perfectly all right then. The week end before he'd had W [*sic*] Gandy over for the week end & they seemed to have had a really good time. Then Mr. & Mrs. Webb came to dinner Tues. & Mrs. Webb had afternoon tea with him Wed: the day she removed.

To Mrs. Clayton, the possibility that Turing had committed suicide seemed inconceivable enough to warrant her offering evidence against it (though not so inconceivable that she felt no need to bring it up). Nonetheless, the result of the inquest, held on June 10, was that Turing had killed himself. It seemed that the apple had been dipped in a cyanide solution.

In the years following his death, many of Turing's friends entered into a sort of conspiracy with his mother to propagate the myth that his death was the result not of suicide but of a scientific experiment gone awry. In cooking up this theory, they pointed to the stock of chemicals (including potassium cyanide) that he kept at his house, as well as his array of scientific equipment. For example, Dr. Greenbaum, the psychoanalyst, wrote to Mrs. Turing,

> There is not the slightest doubt to me that Alan died by an accident. You describe Alans fashion of experimenting so vividly that I can see him pottering about. He was like a child while experimenting not always taking in the observed [illegible] but also testing it with his fingers. . . . [W]hen he died he was never as far away from suicide as there.

Likewise his neighbor, Mrs. Webb, told Mrs. Turing that she found it

> difficult to connect the verdict of the coroner with Alan's behaviour before we left Park Villa on June 3rd. He invited us to dinner on June 1rst and we spent a most delightful evening with him then. I saw him several times during the

> next two days and on the day we moved he invited me in for a cup of tea. He made toast and we had it on the kitchen table. It was such a jolly party, Mrs. Clayton joining us for a cup of tea when she came in. Alan was full of plans for coming to visit us at Styal on his way home from the university in the afternoons, and I cannot believe that he had any idea then of what he was going to do. It must have come upon him quite suddenly.

Hugh Alexander, still in the thick of cryptanalysis, wrote to Mrs. Turing,

> I can confirm what you say about his being in good spirits lately; I had a letter from him about a month before he died saying that he was having treatment, that he felt it was doing him good and that he was in better spirits than he had been [censored].* Because of this I was particularly shocked when I read what had happened and I am very glad to learn that it might well have been an accident.

As late as 1960 Mrs. Turing was still collecting evidence to support her version of events. This last letter came from Turing's former colleague W. T. Jones, now a professor of philosophy at Pomona College in California:

> If I may say so, I think that all of the evidence—both positive and negative—tends to support your views about the circumstances of his death. By "negative" I mean, that I do not think Alan was at all the sort of per-

*Presumably Alexander's letter was censored because of his continuing work for the government in cryptanalysis.

son who would take his own life. By "positive," I mean that he *was* the sort of person who would be careless about (rather, inattentive to) dangerous aspects of the experiments he was conducting.

Interestingly, none of Turing's friends ever seems to have considered, at least in writing, a third possibility (one, admittedly, for which there is no evidence, at present anyway): namely, that the suicide was staged; that the man in the white suit had become—like the hero of Alfred Hitchcock's 1934 film—a man who knew too much.

If he did kill himself, he seems to have thought that he was going somewhere. Remember that in the untitled story, Alec Pryce is an authority on interplanetary travel. In March 1954, a few months before his death, Turing sent Robin Gandy a series of four cryptic postcards. The first was lost. The other three consisted of a list of numbered aphorisms bearing the collective title *Messages from the Unseen World*:

III. The Universe is the interior of the light cone of the creation
IV. Science is a differential Equation. Religion is a Boundary Condition. (sgd) Arthur Stanley*
V. Hyperboloids of wondrous Light
Rolling for aye through Space and Time
Harbour those Waves which somehow might
Play out God's wondrous pantomime
VI. Particles are founts
VII. Charge = e/π ang of character of a 2π rotation

*Arthur Stanley Eddington (1882–1944), mathematical physicist with whom Turing studied at Cambridge.

> VIII. The Exclusion Principle is laid down purely for the benefit of the electrons themselves, who might be corrupted (and become dragons or demons) if allowed to associate too freely

Other mathematicians as great as Turing had ended their lives in madness: Cantor had; also Gödel. Perhaps Turing, too, was becoming delusional near the end, imagining himself rolling through space in a "hyperboloid of wondrous light" known as Pryce's buoy. Or perhaps, as Gandy thought, this was all part of "a new quantum mechanics . . . not intended to be taken very seriously (almost in the 'for amusement only' class), although no doubt he hoped something might turn up in it which could be taken seriously." Or perhaps the new quantum mechanics involved apples, light cones, and spaceships. In *A Mathematician's Apology*, Hardy had written, "No mathematician should ever allow himself to forget that mathematics, more than any other art or science, is a young man's game." Yet Turing, according to Gandy, had not lost his powers; indeed, in the months before his death, he had come up with an upper bound for the Skewes number that was lower than the one that Skewes himself had established. This would have been a significant achievement, had he chosen to publish it. But he did not. He said he didn't want to embarrass Skewes.

The idea of suicide, if it came upon him at all, must have come upon him suddenly. The method, on the other hand, seems to have been in the back of his mind for years. For instance, from Princeton, his friend James Atkins told Andrew Hodges, Turing had once written a letter proposing a suicide method that "involved an apple and electric wiring." He often told his friends that he ate an apple every night before going to bed. And of course, in Cambridge, for weeks after the premiere of *Snow White and the Seven Dwarfs*, he would chant as he walked down the corridors of King's,

Dip the apple in the brew,
Let the sleeping death seep through. . . .

Today the apple continues to fascinate. Much is made of its metaphorical implications. (Apple of death, apple of knowledge—but too much knowledge?) A rumor circulates on the Internet that the apple that is the logo of Apple Computers is meant as a nod to Turing. The company denies any connection; on the contrary, it insists, *its* apple alludes to Newton. But then why has a bite been taken out?

Perhaps what chills us is that in taking his own life, Turing actually chose to camp it up a bit—to invest his departure from a world that had treated him shabbily with some of the gothic, eerie, colorful brilliance of a Disney film. And yet in all the pages I have read about Turing—and there are scores of them—no one has yet mentioned what seems to me the most obvious message. In the fairy tale the apple into which Snow White bites doesn't kill her; it puts her to sleep until the Prince wakes her up with a kiss.

Notes

1: The Man in the White Suit

4 "hounded out of the world": E. M. Forster, *Maurice* (London: Penguin Books, 2000), 32.

5 "Turing believes machines think": Turing Archive, King's College, Cambridge, AMT/D/14a.

8 "Alan certainly had less": Lyn Irvine, preface to Sara Turing, *Alan M. Turing* (Cambridge: W. Heffer, 1959), x.

8 "He never looked right in his clothes": Ibid., xi.

2: Watching the Daisies Grow

9 "Alan was interested in figures": S. Turing, *Alan M. Turing*, 11.

9 "quockling": Ibid., 13.

9 "quite unable to predict": Turing Archive, AMT/K/1/49, Dec. 1936.

10 "a mixture in which the chief ingredient": S. Turing, *Alan M. Turing*, 15.

10 "First you must see that the lite": Ibid., 17.

10 "Turing's fond of the football field": Ibid., 19.

11 "a *quite false impression*": Ibid., 11.

11 "down to something": Ibid., 21.

12 a "world in miniature": E. M. Forster, *The Longest Journey* (London: Penguin Books, n.d.), 157.

12 "not very good": S. Turing, *Alan M. Turing*, 27.

12 "was a cause of satisfaction": Ibid., 27.

13 "No doubt he is very aggravating": Ibid., 29.

13 "What is the *locus* of so and so?": Ibid., 14.

14 "Looking Glass ploy": Andrew Hodges, *Alan Turing: The Enigma* (New York: Walker, 2000), 232.

15 "This room smells of mathematics": Quoted ibid., 29.

15 "Linolite electric strip reflector lamp": Ibid., 56.

15 "the impression that public schools": Ibid., 77.

15 "private locked diary": S. Turing, *Alan M. Turing*, 35.

16 "worshipped the ground": Quoted in Hodges, *Enigma*, 35.

mathematics as a cure for homosexuality: Graham Robb, *Strangers: Homosexual Love in the Nineteenth Century* (New York: W. W. Norton, 2003), 69.

16 "I feel that I shall meet Morcom": Turing Archive, AMT/K/1/20, Feb. 16, 1930.

16 "treasuring with the tenderness": Quoted in Hodges, *Enigma*, 50.

18 "longings": Ibid., 76.

18 "bourgeois, unfinished and stupid": Forster, *Maurice*, 69.

18 "England has always been disinclined": Ibid., 185.

18 "I would rather give a healthy boy": James Douglas, in *Sunday Express*, Aug. 19, 1928; also quoted in Hodges, *Enigma*, 77.

saw *Back to Methuselah*: Hodges, *Enigma*, 74.

not asked to join Cambridge societies: Ibid., 75.

19 "I think I want to talk": Forster, *Longest Journey* 21.

20 His closest friendships: Ibid., 74–76.

21 "At Trinity he would have been": Hodges, *Enigma*, 7.

21 "doubt the axioms": Ibid., 79.

21 "Moore's religion, so to speak": John Maynard Keynes, *Two Memoirs* (London: Rupert Hart-Davis, 1949), 82. I first learned of this fascinating book from Hodges' biography, which quotes from it.

21 “nothing mattered except states of mind”: Ibid., 83.

22 “I have called this faith a religion”: Ibid., 86.

22 “would seem, on Russell’s theory”: G. H. Hardy, “Mathematical Proof,” *Mind*, n.s., 38, 149 (Jan. 1929): 23.

22 “If A was in love with B”: Keynes, Two Memoirs, 86–87.

23 “We entirely repudiated”: Ibid., 97–98.

23 “the fearless uninfluential Cambridge”: Forster, introd. to *Longest Journey*, lxviii.

24 “dowdy, Spartan amateurism”: Hodges, *Enigma*, 69.

24 “Cambridge, I cannot deny”: Forrest Reid, *Private Road* (London: Faber and Faber, 1940), 58.

24 “I pleased one of my lecturers”: Turing Archive, AMT/K/1/23, Jan. 31, 1932.

25 elected a fellow: J. L. Britton, “Remarks on Turing’s Dissertation,” *Pure Mathematics: The Collected Works of A. M. Turing,* ed. J. L. Britton (Amsterdam: North-Holland, 1992), xix.

25 “Turing / Must have been alluring”: Quoted in Hodges, *Enigma*, 94.

26 “I will not be able to pull”: John L. Casti and Werner DePauli, *Gödel: A Life of Logic* (Cambridge: Basic Books, 2000), 117.

28 “The discovery that all mathematics”: Bertrand Russell, “The Study of Mathematics,” in *Contemplation and Action, 1902–14*, ed. Richard A. Rempel, Andrew Brink, and Margaret Moran (London: George Allen and Unwin, 1985), 90.

28 Leibniz’s dream: Martin Davis, *Engines of Logic: Mathematicians and the Origins of the Computer* (New York: W. W. Norton, 2000), 16.

29 “If controversies were to arise”: Russell, “Mathematics and the Metaphysicians,” in *Logicism and the Philosophy of Language: Selections from Frege and Russell*, ed. Arthur Sullivan (Peterborough: Broadview Press, 2003), 224.

29 “every process will represent”: George Boole, *The Mathematical Analysis of Logic: Being an Essay towards a Calculus of Deductive Reasoning* (Cambridge: Macmillan, Barclay, & Macmillan, 1847), 6.

30 “that arithmetic is a branch of logic”: Gottlob Frege, *Grundgesetze der Arithmetik* (Hildesheim: Georg Olms, 1962), translation quoted by Richard G. Heck Jr., in "Julius Caesar and Basic Law V," *http://emerson.fas.harvard.edu/heck/pdf/JuliusCaesarandHP.pdf*

30 "a virulent racist": Davis, *Engines*, 42.

30 "a formal language": Frege, *Begriffsschrift*, in Jean van Heijenoort, *From Frege to Gödel: A Source Book in Mathematical Logic, 1879–1931* (Cambridge: Harvard University Press, 1967), 1.

31 "particular number is not identical": Russell, *Introduction to Mathematical Philosophy* (London: George Allen and Unwin, 1919), 12.

31 "There is just one point": Russell, letter to Frege, in van Heijenoort, *Frege to Gödel*, 124–25.

33 "Your discovery of the contradiction" Frege, letter to Russell, in van Heijenoort, *Frege to Gödel*, pp. 127–28.

33 "a masterpiece discussed": Casti and DePauli, *Gödel*, 43.

34 "the extreme Russellian doctrine": Hardy, "Mathematical Proof," 9.

34 "that mathematics has at its disposal": David Hilbert, "On the Infinite," in van Heijenoort, *Frege to Gödel*, 376.

34 "substantial sciences": Hardy, "Mathematical Proof," 6.

35 "I had better state at once": Ibid., 11–12.

35 "the chessmen, the bat": Ibid.,14–15.

35 "cardinal in Hilbert's logic": Ibid., 15.

36 "because it characterizes": Ernest Nagel and James R. Newman, *Gödel's Proof* (New York: New York University Press, 2001), 28.

36 "conviction of the solvability": Jeremy J. Gray, *The Hilbert Challenge* (Oxford: Oxford University Press, 2000), 248.

36 "there is no such thing as": Constance Reid, *Hilbert* (New York: Springer-Verlag, 1970), 196.

36 "Wir müssen wissen": Gray, *Hilbert Challenge*, 168.

37 "We are convinced that": Reid, *Hilbert*, 188.

37 "Let us consider that": Ibid.

38 "at Berlin University": "Nazi Kultur: The New Heroic Gospel," *Times* (London), Nov. 10, 1933; also quoted in Hodges, *Enigma*, 86.

38 "I am primarily interested": Hardy, "Mathematical Proof," 6.

40 "Let us admit that": Hilbert, "On the Infinite," 375.

40 "a completely satisfactory way": Ibid., 375–76.

40 "The worst that can happen": Hardy, "Mathematical Proof," 5.

Gödel's first public announcement: Robin Gandy, "The Confluence of Ideas in 1936," in *The Universal Turing Machine: A Half-Century Survey*, ed. Rolf Herken, 2nd ed. (Vienna: Springer-Verlag, 1995), 63.

45 "In the highly ingenious work of Gödel": Reid, *Hilbert*, 198.

46 "that proof theory could still": Ibid., 199.

46 "How can one expect": Kurt Gödel, "Russell's Mathematical Logic," in *Collected Works*, vol. 2, ed. Solomon Feferman et al. (Oxford: Oxford University Press, 1990), 140–41.

48 "this did not necessarily mean": Simon Singh, *Fermat's Enigma: The Epic Quest to Solve the World's Greatest Mathematical Problem* (New York: Walker, 1997), 141.

3. The Universal Machine

49 "between two different versions": Egon Börger, Erich Grädel, and Yuri Gurevich, *The Classical Decision Problem* (Berlin: Springer-Verlag, 1997), 4.

50 *Kitab-al-jabr*: Roger Penrose, *The Emperor's New Mind* (Oxford: Oxford University Press, 1989), 40–41.

50 Definition of an algorithm: Ibid., 41–42.

51 "the main problem": Börger et al., *Decision Problem*, 3n.

51 "From the considerations": D. Hilbert and W. Ackermann, *Principles of Mathematical Logic*, ed. Robert E. Luce (New York: Chelsea, 1950), 112.

51 "There is of course no such theorem": Hardy, "Mathematical Proof," 16.

52 "although at present": Börger et al., *Decision Problem*, 5.

52 "a decision procedure might": Letter to the author.

53 "The Hilbert decision-programme": Max Newman, "Royal Society Memoir," in *Mathematical Logic*, ed. R. O. Gandy and

C. E. M. Yates (Amsterdam: Elsevier, 2001), 272.

54 "I remember Turing telling me": Gandy, preface to "On Computable Numbers, with an Application to the *Entscheidungsproblem*," Ibid., 10–11.

55 "We may say most aptly": *Charles Babbage and His Calculating Engine: Selected Writings by Charles Babbage and Others*, ed. Philip Morrison and Emily Morrison (New York: Dover, 1961), 252.

55 Gandy, "The Confluence of Ideas in 1936," 55.

57 "What are the possible processes": Alan Turing, "On Computable Numbers with an Application to the *Entscheidungsproblem*," *Mathematical Logic*, 37.

57 "as the real numbers": Ibid., 18.

57 "it is characteristic of Turing": Hodges, *Turing: A Natural Philosopher* (London: Phoenix, 1997), 8.

57 "the issue of computability": Penrose, *New Mind*, 66.

58 "According to my definition": Turing, "Computable Numbers," 18.

58 "a man in the process": Ibid., 19.

58 "in the machine": Ibid.

59 "At any stage of the motion": Ibid., 20.

59 "The description that he then gave": Newman, "Obituary for Dr. A. M. Turing," *Times* (London), June 16, 1954, 10.

60 "divided into squares": Turing, "Computable Numbers," 38.

61 "lies in the fact that": Ibid., 19.

61 "The difference from our point of view": Ibid., 37–38.

61 "are bound to be": Ibid., 37.

62 "must use successive observations": Ibid., 38.

62 "into 'simple operations'": Ibid.

64 "changes of distribution": Ibid.

64 "immediate recognisability": Ibid., 38–39.

64 "most mathematical papers": Ibid., 39.

65 "simple operations": Ibid.

65 "some of these changes necessarily": Ibid.

65 "We may now construct": Ibid.

72 "The first three symbols on the tape": Ibid., 22.

76 "the convention of writing": Ibid., 23.
77 "a *finitely* described procedure": Stephen C. Kleene, "Turing's Analysis of Computability, and Major Applications of It," *Universal Turing Machine*, 17.
81 "any computable sequence": Turing, "Computable Numbers," 27.
81 "to each computable sequence": Ibid., 29.
82 "We shall avoid confusion": Ibid., 21.
83 "It is possible to invent": Ibid., 29–30.
83 "which will write down": Ibid., 30.
83 "complete configuration": Ibid., 20.
86 "that if *M* can be constructed": Ibid., 30.
86 "at present the machine": Ibid.
87 "It is not altogether obvious": Ibid.
87 "the sequences of letters": Ibid., 30–31.
88 "would be rather more complicated": Penrose, *New Mind*, 55.
89 "convinced themselves that all": Kleene, "Turing's Analysis of Computability," 30.
89 Description number for *U*: Penrose, *New Mind*, 74.
90 "number which is a description number": Turing, "Computable Numbers," 20.
93 "although perfectly sound": Ibid., 34.
93 "gives a certain insight into": Ibid.
94 "we conclude that": Ibid., 35.
96 "By a combination of these processes": Ibid., 36.
97 "In the complete configuration": Ibid., 48.
97 "has the interpretation": Ibid., 47.
97 "There is a general process:" Ibid., 36.

4: God Is Slick

99 "It is difficult to-day": Newman, "Royal Society Memoir," 272.
99 "instruction note": Hodges, *Enigma*, 108.
100 "We have a will": Turing Archive, AMT/C/29, Jan. 31, 1934.
100 "Personally I think that": Ibid.
101 "Then as regards": Ibid.

102 "*Only connect*": Forster, *Howards End* (London: Penguin Books, 1983), 188.

103 "depended, in accordance": Keynes, *Two Memoirs*, 83.

104 "It should probably be remarked": Turing, "Computable Numbers," *Mathematical Logic*, 47.

104 Gödel's attempt to prove the existence of God: Casti and DePauli, *Gödel*, 71–72.

105 "I may add that my objectivistic conception": Hao Wang, *From Mathematics to Philosophy* (London: Routledge and Kegan Paul, 1974), 9.

105 "that Turing machines": Casti and DePauli, *Gödel*, 108.

105 "is analogous to 'the printed book'": Hodges, *Natural Philosopher*, 18.

106 "skeptical of Turing's analysis": Solomon Feferman, "Historical Introduction," *Mathematical Logic*, 3–4.

107 "doing the same things": Turing Archive, AMT/K/1/40, May 29, 1936.

107 "An offprint which you kindly sent me": Quoted in Hodges, *Enigma*, 112.

107 "strong preference for working everything out": Newman, "Royal Society Memoir," 269.

108 "it is almost true to say": Gandy, "Confluence of Ideas in 1936," 78.

108 "a function for which": Casti and DePauli, *Gödel*, 81.

109 "In the following formulation": Emil Post, "Finite Combinatory Processes: Formulation 1," in *The Undecidable: Basic Papers on Undecidable Propositions, Unsolvable Problems and Computable Functions*, ed. Martin Davis (Mineola, N.Y.: Dover, 1993), 289.

110 "like a cross between a panda": Gian-Carlo Rota, "Fine Hall in Its Golden Age: Remembrances of Princeton in the Early Fifties," *http://libweb.princeton.edu/libraries/firestone/rbsc/finding_aids/mathoral/pmcxrota.htm*, 1.

111 "that he often met Church": Interview with Albert Tucker by William Aspray, April 13, 1984, "Mathematical Journals and

Communication," *http://libweb.princeton.edu/libraries/firestone/rbsc/finding_aids/mathoral/pmc32.htm #(PMC32)6*, 5.

111 "toward the end of the tea session": Interview with Albert Tucker by William Aspray, April 11, 1984, "Fine Hall," *http://libweb.princeton.edu/libraries/firestone/rbsc/finding_aids/mathoral/pmc30.htm - (PMC30)9*, 7.

111 "If Weyl says it's obvious": Interview with Stephen C. Kleene and J. Barkley Rosser, by William Aspray, April 26, 1984, *http://libweb.princeton.edu/libraries/firestone/rbsc/finding_aids/mathoral/pmc23.htm*, 6.

111 "began with a ten-minute ceremony": Rota, "Fine Hall," 2.

112 "he was completely oblivious": Interview with Kleene and Rosser, 8.

112 "such a statement": Rota, "Fine Hall," 2.

112 "the remarkable feature": Kleene, "Origins of Recursive Function Theory," *Annals of the History of Computing* 3, no. 1 (Jan. 1981): 62.

112 "for rendering the identification": Ibid., 61.

112 "possibly more convincing": Turing, "Computability and λ-Definability," *Mathematical Logic*, 59.

112 "there were not many others": Interview with Alonzo Church by William Aspray, May 17, 1984, *http://libweb.princeton.edu/libraries/firestone/rbsc/finding_aids/mathoral/pmc05.htm*, 10.

113 "If you don't mind": Ibid., 9.

114 "Of all the ungainly things": S. Turing, *Alan M. Turing*, 51. *Berengaria*: Turing Archive, AMT/K/1/41, Sept. 8, 1936.

115 "The mathematics department here": Ibid., AMT/K/1/42, Oct. 6, 1936.

115 Optimism of Bernays: Gandy, "Confluence of Ideas in 1936," 59.

115 "he was very standoffish": Turing Archive, AMT/K/1/43, Oct. 14, 1936.

117 "not always understood": *A Princeton Companion*, *http://etc.princeton.edu/CampusWWW/Companion/veblenoswald.html.*

118 "honorary member of the clique": Interview with Shaun Wylie by Frederik Nebeker, June 21, 1985,*http://ibweb.princeton.edu/libraries/firestone/rbsc/finding_aids/mathoral/pmc45.htm*, 10.

119 "talking shop": Turing Archive, AMT/K/1/43, Oct. 14, 1936.

119 "Though prepared to find": S. Turing, *Alan M. Turing*, 52.

120 "I am sending you some cuttings": Turing Archive, AMT/K/1/46, Nov. 22, 1936.

120 "horrified at the way people": Ibid., AMT/K/1/48, Dec. 3, 1936.

120 "I believe the government": Ibid., AMT/K/1/51, Jan. 1, 1937.

121 "glad that the Royal Family": Ibid., AMT/K/1/59, May 19, 1937.

121 "Church had me out to dinner": Ibid., AMT/K/1/43, Oct. 14, 1936.

121 "Yes, I forgot about him": Church/Aspray, 10.

122 "I have had two letters": Turing Archive, AMT/K/1/56, Feb. 22, 1937.

122 von Neumann's talk with Gödel: Casti and DePauli, *Gödel*, 50.

122 "was one of the first": Feferman, "Turing in the Land of O(z)," *Universal Turing Machine*, 113.

123 "You had von Neumann": Interview with Joseph Daly and Churchill Eisenhart by William Aspray, July 10, 1984, *http://libweb.princeton.edu/libraries/firestone/rbsc/finding_aids/mathoral/pmc07.htm*, 4.

124 "Maurice is much more conscious": Turing Archive, AMT/K/1/57, March 29, 1937.

125 "As a matter of fact": Alonzo Church, review of "On Computable Numbers, with an Application to the *Entscheidungsproblem*." *Journal of Symbolic Logic* 2, no. 1 (March 1937): 43.

125 "actually easier to work with": Aspray/Kleene/Rosser, PMC23, 10.

126 "One should have a reputation": Turing Archive, AMT/K/1/51, Jan. 1, 1937.

127 "I went to the Eisenharts": Ibid., AMT/K/1/56, Feb. 22, 1937.

127 "a rich man": Ibid., AMT/K/1/59, May 19, 1937.

130 Hardy's pessimism: Marcus Du Sautoy, *The Music of the Primes: Searching to Solve the Greatest Mystery in Mathematics* (New York: HarperCollins, 2003), 188.

131 "The number of protons": Hardy, *Ramanujan* (Cambridge: Cambridge University Press, 1940), 17.

131 Lehman's method: Feferman, "Turing in the Land of O(z)," 110.

132 "a generation": Hodges, *Enigma*, 118.

132 "the 'real' mathematics": Hardy, *A Mathematician's Apology* (Cambridge: Cambridge University Press, 1967), 119–20.

132 "no one has yet discovered": Ibid., 140.

132 "gentle and clean": Ibid., 121.

133 "You have often asked me": Turing Archive, AMT/K/1/43, Oct. 14, 1936.

134 "multiply the number": Quoted in Hodges, *Enigma*, 138.

136 "regarded as thoroughly unsatisfactory": Kleene, "General Recursive Function," 59.

136 "only after Turing's formulation": Ibid., 61.

137 "giving an absolute definition": Gödel, "Remarks before the Princeton Bicentennial Conference on Problems in Mathematics, 1946," in Davis, *Undecidable*, 84.

137 "due to A. M. Turing's": Ibid., 71.

137 Gödel's dissatisfaction: Feferman, "Historical Introduction," 5–6.

138 "pure mathematics is the subject": Casti and DePauli, *Gödel*, 28.

138 "distanc[ing] himself from": Ibid., 71.

5: The Tender Peel

139 "a part of his mind": Hodges, *Enigma*, 148–49.

140 Turing went to see *Snow White*: Ibid., 149.

141 attended another training course: Ibid., 151.

141 "Apparatus would be": Quoted ibid., 155.

142 design of Liverpool machine: Du Sautoy, *Music of the Primes*, 188.

142 precision-cut gear wheels: Hodges, *Enigma*, 156.

143 two courses with the same name: Ibid., 152.

144 "to be an elderly man": Norman Malcolm, *Ludwig Wittgenstein: A Memoir* (London: Oxford University Press, 1958), 23.

145 "were given without preparation": Ibid., 24.

145 "He always wore": Ibid., 24–25.

145 "were austerely furnished": Ibid., 25.

146 "One had to be brave": Ibid., 25–26.

146 "I might as well talk": Ibid., 26–27.

146 "Wittgenstein applied his own": Casti and DePauli, *Gödel*, 71–72.

146 "Suppose I say to Turing": *Wittgenstein's Lectures on the Foundations of Mathematics, Cambridge 1939*, ed. Cora Diamond (Chicago: University of Chicago Press, 1989), 20.

147 "Don't treat your common sense": Ibid., 68.

147 "tempt": Ibid., 139.

147 "I understand but": Ibid., 67

148 "We say of a proof that": Ibid., 199.

148 "Professor Hardy says": Ibid., 138–39.

149 "What is counting?": Ibid., 115.

149 "that whenever numerals": Ibid., 31.

149 "You might call this figure": Ibid., 36–37.

150 "The ordinary meanings of": Ibid., 37.

151 "One could make this comparison": Ibid., 96–97.

152 "If a man says": Ibid., 206–7.

152 "I may give you the rules": Ibid., 210–11.

153 "a logical system": Ibid., 212.

153 "practical things may go wrong": Ibid., 216.

153 "The question is": Ibid., 217.

154 "But how do you know that": Ibid., 218.

155 "Before we stop": Ibid., 219–20.

156 "out of the paradise": Ibid., 103.

156 "Suppose I am a general and I receive": Ibid., 201.

157 "Suppose I am a general and I give": Ibid., 212.

157 "a sort of Victorian mock-Tudor": Stephen Budiansky, *Battle of Wits: The Complete Story of Codebreaking in World War II* (New York: Touchstone, 2000), 118.

157 "even to the untrained eye": David Russo, "Architecture and the Architect," *http://www.utdallas.edu/~dtr021000/cse4352/architects.doc.*

159 alter the orders of the keyword: Budiansky, *Battle of Wits*, 67.

165 "Sphinx of the Wireless": Singh, *The Code Book: The Secret History of Codes and Codebreaking* (London: Fourth Estate, 1999), 138.

171 100,391,791,500 further permutations: Ibid., 136.

incorporating the indicator code: Hodges, *Enigma*, 164.

176 "Make sure they have": Quoted ibid., 221.

178 "For centuries": Singh, *Code Book*, 149.

179 "keine Zusätze": Turing, Excerpts from the "Enigma Paper," *Mathematical Logic*, 230–31.

184 the gunner on an English ship: Budiansky, *Battle of Wits*, 157.

186 "the fundamental mathematical insight": Ibid., 131.

186 "*these* contradictions": Hodges, *Enigma*, 183–84.

obsessive counting of tire revolutions: Ibid., 209.

187 "pompousness or officialdom": Quoted ibid., 204.

187 "much more conscious": Turing Archive, AMT/K/1/57, March 29, 1937.

189 "we might have lost": Singh, *Code Book*, 176.

6: The Electronic Athlete

192 Subsequent Robinsons: Hodges, *Enigma*, 267n.

193 a stranger propositioned him: Ibid., 249.

196 "found the idea of": Irvine, preface to S. Turing, *Alan M. Turing*, xii.

196 "seemed to think it": Hodges, *Enigma*, 284.

196 "If it had ended unhappily": Forster, *Maurice*, 218.

197 "had hoped for": Ibid., 221–22.

197 "Sometimes you're sitting": Quoted in Hodges, *Enigma*, 373.

198 "the possible adaptation": Quoted ibid., 306.

199 17,468 vacuum tubes: Mary Bellis, "Inventors of the Modern Computer," *http://inventors.about.com/library/weekly/aa060298.htm*, 1.

199 "With the advent of": Martin H. Welk, "The ENIAC Story," *http://ftp.arl.mil/~mike/comphist/eniac-story.html*, 1.

200 "screwdriver interference": Turing, "Intelligent Machinery," in *Mechanical Intelligence*, ed. D. C. Ince (Amsterdam: North-Holland, 1992), 115.

201 "While it appeared that": John von Neumann, "First Draft of a Report on the EDVAC," June 30, 1945, 3.

202 "tackle whole problems": Turing, "Proposal for Development in the Mathematics Division of an Automatic Computing Engine (ACE)," *Mechanical Intelligence*, 1.

202 "There will positively be no": Ibid., 2.

203 "Possibly it might still": Turing, "Lecture to the London Mathematical Society on 20 February 1947," *Mechanical Intelligence*, 104.

203 if not "infinite": Ibid., 88.

203 "desirable feature": Ibid., 89.

204 "form of memory": Ibid., 88.

204 "really fast machine": Ibid., 89.

204 "cuts out the sign": *Wittgenstein's Lectures*, 20.

204 "The machine interprets": Turing, "Lecture to the London Mathematical Society," 103.

205 "I would say that fair play": Ibid., 104–5.

206 "To continue my plea": Ibid., 105.

207 "the calculator itself": Ibid., 102.

208 "That it is electronic": Ibid., 87.

209 "Construction of range tables": Turing, "Report on the ACE," *Mechanical Intelligence*, 20–22.

209 "be made to do any job": Turing, "Lecture to the London Mathematical Society," 87.

209 "To perform the various logical operations": Turing, "Report on the ACE," 8.

209 magnetic wire: Turing, "Lecture to the London Mathematical Society," 89.

210 "much the most hopeful": Ibid., 84.

210 "It should be possible": Turing, "Report on the ACE," 20.

210 "a standard instruction": Ibid., 17.

210 "have to be made up by": Ibid., 25.

211 "grasp the principle": Hodges, *Enigma*, 335.

211 "The most that Alan": S. Turing, *Alan M. Turing*, 70.

212 "comprised men from": Ibid., 111.

212 "about twelve years ago": Quoted ibid., 79.

213 "evening paper": Ibid., 80.

213 "be able quite easily": Quoted ibid., 81.

213 "power of judgment": Quoted ibid.
215 "I have read Wilkes' proposal": Quoted in Hodges, *Enigma*, 352.
215 "intended primarily": Quoted ibid., 353.
215 "very opinionated": Quoted ibid., 353.
215 "minimalist ideas": Davis, *Engines*, 189.
216 "the actual size of": Quoted in Hodges, *Enigma*, 408.
216 "to be very much alone": Davis, *Engines*, 192.
216 "Virtually all computers": Quoted ibid., 193.
217 "disappointed with": S. Turing, *Alan M. Turing*, 86–87.
218 "was not a particularly good": Quoted ibid., 87.
218 "to the similarities": Ibid., 88.
219 "With this store available": "Max Newman and the Mark I," *http://www.computer50.org/mark1/newman.html*, 2.
220 "mathematical problems": Hodges, *Enigma*, 341.
220 looked like a beetle: Ibid., 375.
221 use "as much of": Quoted ibid., 375.
222 "machinery might be": Turing, "Intelligent Machinery," 107.
222 "an unwillingness": Ibid., 107–8.
222 "being purely emotional": Ibid., 108.
223 "can go on through": Ibid.
223 "It is related that": Ibid., 108–9.
224 "that intelligence in machinery": Ibid., 107.
224 "the view that the credit": Ibid., 109.
224 "form a continuous": Ibid.
224 "intended to produce": Ibid.
224 "discrete controlling": Ibid., 110.
225 "A great positive reason": Ibid. 116–17.
226 "take a man as a whole": Ibid., 117.
226 five possible applications: Ibid.
227 "The training of the human child": Ibid., 121.
228 "If the untrained infant's mind": Ibid., 125.
228 "the machine very soon": Ibid., 123.
229 "Certainly the nerve": Ibid., 117.
229 "is determined as much": Ibid., 127.

7: The Imitation Game

231 "late lavatorial": Quoted in Hodges, *Enigma*, 391.

231 "carpeted in": W. G. Sebald, *The Emigrants*, trans. Michael Hulse (New York: New Directions, 1996), 151.

232 "to a number": Turing, "Programmer's Handbook (2nd Edition) for the Manchester Electronic Computer Mark II," *http://www.computer50.org/kgill/mark1/program.html*, 3.

232 "saddled users": Martin Campbell-Kelly, "Turing's Papers on Programming," *Mathematical Logic*, 244.

232 "Because zero was": Ibid., 245.

233 "bizarre in the extreme": Ibid.

234 Lucas's method: Hodges, *Enigma*, 398.

234 "As every vehicle": Quoted ibid., 402; from an interview with Martin Campbell-Kelly.

236 "Not until a machine": *British Medical Journal*, June 25, 1949; quoted in Hodges, *Enigma*, 405.

237 "create concepts": "No Mind for Mechanical Man," *Times* (London), June 10, 1949, 2.

237 "This is only a foretaste": "The Mechanical Brain," *Times* (London), June 11, 1949, 4.

238 "Isn't that just like": S. Turing, *Alan M. Turing*, 91.

238 "The university was": "The Mechanical Brain," 4.

239 "the rather mysterious description": "The Mechanical Brain: Successful Use of Memory-Storage," *Times* (London), June 14, 1949, 5.

239 "responsible scientists": Illtyd Trethowan, letter to the editor, *Times* (London), June 14, 1949, 5.

239 "There would be plenty to do": "Intelligent Machinery: A Heretical Theory," in S. Turing, *Alan M. Turing*, 133–34.

240 "those who have never loved": "Umbrage of Parrots," *Times* (London), June 16, 1949, 5.

241 "I propose to consider": Turing, "Computing Machinery and Intelligence," 133.

241 “is played with three people”: Ibid., 133–34.
243 “Turing’s gender-guessing”: Hodges, *Natural Philosopher*, 38.
243 “The new problem”: Turing, “Computing Machinery and Intelligence,” 134–35.
244 “It might be urged”: Ibid., 135.
244 “I put it to you”: Turing Archive, AMT/B/6, 6.
245 “we wish to exclude”: Turing, “Computing Machinery and Intelligence,” 135–36.
245 “Scudder, why do you think”: Forster, *Maurice*, 194.
246 “domestic analogy”: Turing, “Computing Machinery and Intelligence,” 138.
246 “I believe that in about fifty years’ time”: Ibid, 142.
247 “the theological objection”: Ibid., 143.
248 “In attempting to construct”: Ibid.
248 “The consequences of”: Ibid., 144.
249 “the appropriate critical question”: Ibid., 145.
250 “This argument appears”: Ibid., 146.
250 “sure that Professor Jefferson”: Ibid.
250 “to discover whether someone”: Ibid.
251 “than be forced into”: Ibid., 147.
251 “Be kind, resourceful, beautiful”: Ibid., 147–48 (with some minor punctuation changes).
252 “There are, however, special remarks”: Ibid., 148.
252 “Do you eat oysters?”: Quoted in *http://www.outsmartmagazine.com/issue/i06-02/tonycurtis.php*.
253 “The claim that”: Turing, “Computing Machinery and Intelligence,” 148.
254 “Errors of functioning”: Ibid., 149.
254 “that a machine cannot”: Ibid.
254 “a computer can do *whatever*”: Ibid., 150.
254 “if we adhere to”: Ibid., 151.
254 “If each man had”: Ibid., 152.
255 “I have set up”: Ibid., 153.
255 “Unfortunately the statistical evidence”: Ibid., 153.
255 “Let us play”: Ibid.

256 “Then it will be natural”: Ibid., 153–54.
256 “presuppose any feelings”: Ibid., 157.
257 “the use of punishments”: Ibid.
257 “Instead of trying”: Ibid., 156.
257 “It will not, for instance”: Ibid.
258 “the imperatives that”: Ibid., 158.
258 “the rules which get changed”: Ibid.
258 “will eventually compete”: Ibid., 160.
259 “God Save the King”: Hodges, *Enigma*, 447.
260 “a negligible advance”: Turing, “Some Calculations of the Riemann Zeta-Function,” *Pure Mathematics*, 97.

8. Pryce’s Buoy

261 “I am so glad”: S. Turing, *Alan M. Turing*, 103.
261 “high emotional content”: Turing Archive, AMT/B/6, 26.
262 “the intervention of”: Ibid., 23.
262 “When the work”: Ibid., 33.
262 “a computing machine”: Ibid., 20.
262 “necessary for the machine”: Ibid., 28.
262 “more interested in curbing”: Ibid., 29.
263 “until he saw it touch”: Hodges, *Enigma*, 452.
263 “That old slow coach”: Turing Archive, AMT/B/6, 36.
264 “not interested in the fact”: Turing Archive, AMT/B/6, 5.
264 “Turing had arrived”: Newman, “Royal Society Memoir,” 278.
265 “he shared many jokes”: S. Turing, *Alan M. Turing*, 92.
meeting Arnold Murray: Hodges, *Enigma*, 450.
£50 worth of his belongings: Ibid., 454.
267 “do his worst”: Ibid., 455.
267 “Mr. Hall—you reckernize”: Forster, *Maurice*, 193.
267 “By God, if you’d spilt”: Ibid., 196.
268 “knew all about”: Hodges, *Enigma*, 456.
268 “I don’t think I really”: Turing Archive, AMT/D/14a, 1952.
270 “I have a delightful story”: Turing Archive, AMT/D/14a, 1953.
271 “Alec always felt”: Turing Archive, AMT/A/13, undated.

272 "He didn't care to wear": Ibid.
272 "had been out of work": Ibid.
272 "very hungry and rather cold": Ibid.
273 "That chap who was walking round": Ibid.
273 "He didn't smoke": Ibid.
274 "'Don't mind if I do' ": Ibid.
274 "the door opened for you": Ibid.
275 "You will by now have heard": Turing Archive, AMT/A/15, June 6, 1954.
276 "There is not the slightest doubt": Turing Archive, AMT/A/16, May 1, 1955.
276 "difficult to connect": Turing Archive, AMT/A/17, June 13, 1954.
277 "I can confirm": Turing Archive, AMT/A/17, Aug. 18, 1954.
277 "If I may say so": Turing Archive, AMT/A/23, Sept. 24, 1960.
Messages from the Unseen World: "The Letter Written by Robin Gandy to Max Newman in June 1954," *Mathematical Logic*, 267.
279 "a new quantum mechanics": Ibid., 266.
279 "No mathematician should ever": Hardy, *Apology*, 70.
279 "involved an apple": Hodges, *Enigma*, 129.

Further Reading

For any reader interested in learning more about Alan Turing, there is nowhere better to begin than with Andrew Hodges' *Alan Turing: The Enigma* (Walker, 2000). This fine biography is at once shrewd, sensitive, and exhaustive—the sort of book that makes other books possible.

The most important of Turing's papers—including "Computable Numbers" and "Mechanical Intelligence"—have been collected in *The Essential Turing: The Ideas That Gave Birth to the Computer Age*, edited by B. Jack Copeland and published by Oxford University Press on the occasion of what would have been the mathematician's ninetieth birthday. Turing's complete writings can be found in the four-volume *Collected Works of A. M. Turing*, published by North-Holland. Of particular interest are the volumes entitled *Mathematical Logic* (2001) and *Mechanical Intelligence* (1992). Turing's letters, as well as letters to him and the drafts of some of his papers, are stored in the archives at King's College, Cambridge. (I am indebted to Dr. Rosamund Moad for making these documents available to me.) A number of these can be consulted online at *http://www.turing archive.org*.

Surprisingly few books take on Turing as their exclusive subject. Concise introductions to Turing's ideas can be found in Hodges' *Turing: A Natural Philosopher* (Phoenix, 1997) and Jon Agar's *Turing and the*

Universal Machine (Icon, 2001). *Alan Turing: Life and Legacy of a Great Thinker*, edited by Christof Teuscher (Springer-Verlag, 2004), brings together essays (and a play) on Turing by, among others, Hodges, Martin Davis, Daniel Dennett, and Douglas Hofstadter.

Many of the other primary texts to which I refer have been collected in three useful omnibus volumes: *The Undecidable: Basic Papers on Undecidable Propositions, Unsolvable Problems and Computable Functions*, edited by Martin Davis (Dover, 1993); *From Frege to Gödel: A Source Book in Mathematical Logic, 1879–1931*, edited by Jean van Heijenoort (Harvard University Press, 1967); and *The Universal Turing Machine: A Half-Century Survey*, edited by Rolf Herkin (Springer-Verlag, 1995). In addition, the interviews with Princeton mathematicians cited in chapter 4 can be consulted online at *http://infoshare1.princeton.edu/libraries/firestone/rbsc/finding_aids/mathoral/math.html.* Finally, many documents from the period during which Turing worked at Manchester can be viewed at *http://www. computer50.org.*

For those interested in learning about the prehistory of computers, Martin Davis's *Engines of Logic: Mathematicians and the Origins of the Computer* (Norton, 2000) provides a lucid and thorough introduction. I would also like to recommend David Berlinski's *The Advent of the Algorithm: The Three-Hundred Year Journey from an Idea to the Computer (Harvest, 2001)*. Good overviews of Gödel's work include *Gödel's Proof*, by Ernest Nagel and James R. Newman (New York University Press, 2001): *Gödel: A Life of Logic*, by John L. Casti and Werner DePauli (Basic Books, 2000), and Rebecca Goldstein's *Incompleteness: The Proof and Paradox of Kurt Gödel* (Norton/Atlas, 2005). A diverting riff on Georg Cantor's work on infinity, including the Diagonal argument, can be found in David Foster Wallace's *Everything and More: A Compact History of ∞* (Norton/Atlas, 2003). Jeremy J. Gray's *The Hilbert Challenge* (Oxford University Press, 2000) offers a fascinating overview of Hilbert's "program," while Constance Reid's *Hilbert* (Springer-Verlag, 1970) is a surprisingly moving account of this great mathematician's life and work.

The year 2003 saw the publication of three equally readable, yet utterly distinct, books on the Riemann hypothesis: John Derbyshire's *Prime Obsession: Bernhard Riemann and the Greatest Unsolved Problem in*

Mathematics (Joseph Henry Press), Marcus du Sautoy's *The Music of the Primes: Searching to Solve the Greatest Mystery in Mathematics* (HarperCollins), and Karl Sabbagh's *The Riemann Hypothesis: The Greatest Unsolved Problem in Mathematics* (Farrar, Straus, Giroux). These were joined, in 2005, by Dan Rockmore's witty and engaging *Stalking the Riemann Hypothesis: The Quest to Find the Hidden Law of Prime Numbers* (Pantheon, 2005). This last book, cleverly, puts the prime page numbers in boldface. Of the numerous accounts in print of Turing's work at Bletchley Park during World War II, I would particularly recommend the one in Stephen Budiansky's *Battle of Wits: The Complete Story of Codebreaking in World War II* (Touchstone, 2000). A "virtual tour" of Bletchley Park can be taken at *http://www.bletchleypark.org.uk*.

Since Turing's death, numerous books and essays have appeared that interrogate, challenge, and extend his arguments. Of these, the most compelling—at least for me—are Roger Penrose's *The Emperor's New Mind: Concerning Computers, Minds and the Laws of Physics* (Oxford University Press, 1999) and John Searle's *Minds, Brains, and Science* (Harvard University Press, 1984), which outlines the now infamous "thought experiment" of the Chinese room.

Lastly, the Alan Turing homepage at *http://www.turing.org.uk/*, as maintained by Andrew Hodges, remains a touchstone for anyone interested in the life and work of this great mathematician.

I am indebted to Jesse Cohen and Prabhakar Ragde for their help in the preparation of this book.

Index

Page numbers in *italics* refer to illustrations.